Lecture Notes in Mathematics

2096

For further volumes:
http://www.springer.com/series/304

Stanislav Hencl • Pekka Koskela

Lectures on Mappings of Finite Distortion

 Springer

Stanislav Hencl
Department of Mathematical Analysis
Charles University
Prague 8, Czech Republic

Pekka Koskela
Department of Mathematics and Statistics
University of Jyväskylä
Jyväskylä, Finland

ISBN 978-3-319-03172-9   ISBN 978-3-319-03173-6 (eBook)
DOI 10.1007/978-3-319-03173-6
Springer Cham Heidelberg New York Dordrecht London

Lecture Notes in Mathematics ISSN print edition: 0075-8434
ISSN electronic edition: 1617-9692

Library of Congress Control Number: 2013957804

Mathematics Subject Classification (2010): 30C65, 46E35, 26B10

Printed on acid-free paper

Springer is part of Springer Science+Business Media (www.springer.com)

Preface

This material is based on the graduate level courses that the authors have given at the University of Michigan, University of Jyväskylä, and Charles University in Prague and on short courses by the authors at summer schools in Ischia and at the de Giorgi Center in Pisa. We thank the participants of these courses for their questions that have shaped the contents and for pointing out a number of mistakes in previous versions.

In order to make the topic accessible to a beginning graduate student, we have included a great number of details, especially in the first four chapters that can form a basis for a graduate course. Additionally, we have recorded all the necessary background material from real analysis and from the theory of Sobolev spaces that is not necessarily covered in undergraduate studies or in the basic graduate level real analysis courses. The later chapters partially cover very recent research, not included in the research monographs [4, 67] that we recommend for further reading.

Finally, we wish to thank our colleagues and graduate students, especially Sita Benedict, Daniel Campbell, Sebastiano Nicolussi Golo, Ville Kirsilä, Luděk Kleprlík, Jan Malý, Gaven Martin, Jani Onninen, Kai Rajala, Eero Saksman, Ville Tengvall, and Aleksandra Zapadinskaya, for their pointed comments. The authors acknowledge the support of the following grants: ERC CZ grant LL1203 of the Czech Ministry of Education (S.H) and Academy of Finland grant number 131477 (P.K.).

Prague, Czech Republic   Stanislav Hencl
Jyväskylä, Finland   Pekka Koskela
October 2013

Contents

Notation

$\mathbf{N}, \mathbf{Z}, \mathbf{R}$	Positive integers, integers, real numbers
(a, b)	Open interval in $\mathbf{R}$ for $a, b \in \mathbf{R}$, $a < b$
$[a, b]$	Closed interval in $\mathbf{R}$ for $a, b \in \mathbf{R}$, $a < b$
$\mathbf{R}^n$	n-dimensional Euclidean space
$|x|$	The Euclidean norm of a vector $x \in \mathbf{R}^n$
$\|x\|$	In Chap. 4 we use this notation for maximum norm of $x \in \mathbf{R}^n$
$\langle u, v \rangle$	Usual inner product of vectors $u, v \in \mathbf{R}^n$, i.e. $\langle u, v \rangle = \sum_{i=1}^{n} u_i v_i$
$u \otimes v$	Tensor product of vectors $u, v \in \mathbf{R}^n$, i.e. $n \times n$ matrix $\{u_i v_j\}_{i,j=1}^{n}$
$|E|$	Operator norm of the matrix E, i.e. $\sup\{|Ex| : |x| \leq 1\}$
I	Identity matrix, i.e. 1 on the diagonal and 0 otherwise
$\operatorname{adj} E$	Adjoint matrix of the matrix E, i.e. $E \operatorname{adj} E = I \det E$
$B(c, r)$	Open ball centered at $c \in \mathbf{R}^n$ with radius $r > 0$, i.e. $\{x \in \mathbf{R}^n : |x - c| < r\}$
$Q(c, r)$	Open cube centered at $c \in \mathbf{R}^n$ with radius $r > 0$, i.e. $\{x \in \mathbf{R}^n : \|x - c\| < r\}$
$t B(c, r)$	Inflated ball for $t > 0$, i.e. $B(c, tr)$
$t Q(c, r)$	Inflated cube for $t > 0$, i.e. $Q(c, tr)$
$S^{n-1}(c, r)$	Sphere centered at $c \in \mathbf{R}^n$ with radius $r > 0$, i.e. $\{x \in \mathbf{R}^n : |x - c| = r\}$
ω_{n-1}	$(n-1)$-dimensional measure of $S^{n-1}(0, 1)$
$[x, y]$	Line segment connecting $x, y \in \mathbf{R}^n$
$\operatorname{dist}(x, A)$	Distance of a point $x \in \mathbf{R}^n$ to a set $A \subset \mathbf{R}^n$
$\operatorname{dist}(A, B)$	Distance of two sets $A, B \subset \mathbf{R}^n$
$\overline{A}$	Closure of a set $A \subset \mathbf{R}^n$
∂A	Boundary of a set $A \subset \mathbf{R}^n$
$\operatorname{diam} A$	Diameter of a set $A \subset \mathbf{R}^n$, $\operatorname{diam} A = \sup\{|x - y| : x, y \in A\}$
Ω	By Ω we always denote an open subset of $\mathbf{R}^n$
$A \subset\subset \Omega$	Set A is compactly contained in Ω, i.e. $\overline{A} \subset \Omega$ and $\overline{A}$ is compact
$|A|$ or $\mathscr{L}_n(A)$	n-dimensional Lebesgue measure of a measurable set A

χ_A	Characteristic function of a set A, i.e. 1 on A and 0 otherwise
$\#A$	Cardinality of the set A, i.e. the number of the elements in A
sgn	Sign function, i.e. sgn $t = 1$ for $t > 0$, sgn $t = -1$ for $t < 0$ and sgn $0 = 0$
$f^+(x)$	Nonnegative part of function f, i.e. $\max\{f(x), 0\}$
spt f	Support of a function f, spt $f = \overline{\{x : f(x) \neq 0\}}$
$L^p(\Omega)$	Lebesgue space—see Appendix for the definition
$\|f\|_p, \|f\|_{L^p}$	The L^p norm of function f
$L^p_{\text{loc}}(\Omega)$	Local Lebesgue space—see Appendix for the definition
$L^n \log^\alpha L(\Omega)$	Zygmund space—see Definition 2.6
$WL^n \log^\alpha L(\Omega)$	Sobolev Zygmund space—see Definition 2.6
$W^{1,p}(\Omega)$	Sobolev space—see Appendix for the definition
$W^{1,p}_{\text{loc}}(\Omega)$	Local Sobolev space—see Appendix for the definition
$W^{1,p}_0(\Omega)$	Sobolev space with zero boundary values—see Appendix for the definition
$BV(\Omega)$	The space of functions with bounded variation, see Definition 5.1
∇f	Classical derivative (gradient) of function $f : \Omega \to \mathbf{R}$ or mapping $f : \Omega \to \mathbf{R}^n$
Df	Weak derivative of function or mapping f—see Definition A.13
$J_f(x)$	Jacobian, i.e. the determinant of $Df(x)$ for $f = (f_1, \ldots, f_n) : \Omega \to \mathbf{R}^n$. Sometimes we use $J(f_1, f_2, \ldots, f_n)(x)$ to point out the components of f
K_f or K	Distortion of function f, see Definition 1.11
K_I	Inner distortion function, see Sect. 7.1
$\mathscr{J}_f$	Distributional Jacobian, see Sect. 2.2
$\deg(C, f, U)$	Topological degree of f on a set C with respect to U, see Sect. 3.2
$N(f, \Omega, y)$	Number of preimages of point y in Ω under f
C, C_C	Continuous functions, continuous functions with compact support
$C^1 (C^2)$	The class of functions with continuous first order (second order) derivatives
$C^\infty_C(\Omega)$	The class of compactly supported (spt $f \subset\subset \Omega$) functions with derivatives of all orders
$C^\infty_0(\Omega)$	Functions whose extension by 0 to $\mathbf{R}^n \setminus \Omega$ belong to $C^\infty(\mathbf{R}^n)$
$f_A = \fint_A f$	Integral average of $f : A \to \mathbf{R}$ defined as $\frac{1}{|A|} \int_A f(x) \, dx$
$\text{osc}_B f$	Oscillation of f on a set B, i.e. diameter of the image $\text{osc}_B f = \text{diam } f(B)$
Mf	Maximal operator of f, see Sect. 7.3
$\mathscr{H}^k$	The k-dimensional Hausdorff measure
$\mathscr{H}^k_\varepsilon$	Set functions in the definition of $\mathscr{H}^k$, i.e. $\mathscr{H}^k(A) = \lim_{\varepsilon \to 0+} \mathscr{H}^k_\varepsilon(A)$

$\int_{S^{n-1}(c,t)} f$	Integral of f with respect to the surface measure (constant multiple of $\mathcal{H}^{n-1}$)
C	We use the usual convention that C denotes a generic positive constant whose exact value may change at each occurrence
$a \approx b$	Means that $a \leq Cb$ and $b \leq Ca$

Chapter 1
Introduction

Abstract In this chapter we introduce mappings of finite distortion as a generalization of mappings of bounded distortion. We show that this class is natural for the regularity of inverse mappings and for models in nonlinear elasticity.

In 1981, Ball [9] established an invertibility property for Sobolev mappings and used this to show that the solution of the displacement boundary value problem of nonlinear elastostatics does not violate the principle of interpenetration of matter. He posed an open problem that in a simplified form asks the following:

Suppose that a continuous mapping $f : B(0, 1) \to B(0, 1)$, in $\mathbf{R}^n$ for some $n \geq 2$, belongs to the Sobolev class $W^{1,n}(B(0, 1), \mathbf{R}^n)$, satisfies $J_f(x) := \det Df(x) > 0$ almost everywhere, f is a homeomorphism from $B(0, 1) \setminus \overline{B(0, 1 - \delta)}$ onto $B(0, 1) \setminus f(\overline{B(0, 1 - \delta)})$, and f additionally satisfies

$$\int_{B(0,1)} |(Df(x))^{-1}|^n J_f(x)\, dx < \infty. \tag{1.1}$$

Does it follow that the mapping f has an inverse $f^{-1} : B(0, 1) \to B(0, 1)$ with $f^{-1} \in W^{1,n}(B(0, 1), \mathbf{R}^n)$? Above and in what follows, $|A|$ for a matrix A refers to the operator norm of A.

The original idea of Ball [8,9] was to establish a class of mappings that can serve as a class of deformations in nonlinear elasticity. Nowadays the whole theory is very rich and we recommend the monographs [4, 67] for history, references and further motivation.

We can view a domain $\Omega \subset \mathbf{R}^n$ as a solid body in space and our mapping $f : \Omega \to \mathbf{R}^n$ as a deformation of the body Ω to $f(\Omega)$. There are several natural questions one can ask.

- Is f continuous? (Does the material break or are there any cavities created during the deformation?)

S. Hencl and P. Koskela, *Lectures on Mappings of Finite Distortion*, Lecture Notes in Mathematics 2096, DOI 10.1007/978-3-319-03173-6_1,
© Springer International Publishing Switzerland 2014

- Does f map sets of measure zero to sets of measure zero? (Is new material created from "nothing"? Is some material "lost" during the deformation?)
- Is the mapping one-to-one? Does the inverse map f^{-1} exist? (Is there no interpenetration of the matter, i.e. can we map two parts of the body to the same place? Can we deform the body back to its original state?)
- What are the properties of f^{-1}? (Is the reverse deformation reasonable?)

In the planar setting, basic linear algebra transforms (1.1) to the requirements that $J_f(x) > 0$ almost everywhere and that the quantity

$$K_f(x) := \frac{|Df(x)|^2}{J_f(x)}$$

is integrable over $B(0, 1)$. The latter condition is a relaxation of the definition of quasiregularity (or quasiconformality) that requires that infinitesimal circles be mapped to infinitesimal ellipses whose eccentricities $K_f(x)$ are uniformly bounded. Thus the study of mappings as in (1.1) can be viewed as a generalization of the study of quasiregular mappings, also called mappings of bounded distortion. In a sense, one relaxes the boundedness of the distortion to integrability of the distortion. We will give the relevant definitions in Sect. 1.2 below. Mappings of bounded distortion are continuous, map sets of measure zero to sets of measure zero, and they are either constant or locally bounded to one. Also, a mapping of bounded distortion that is injective close to the boundary is necessarily a homeomorphism and the inverse is also of bounded distortion. Thus this class of mappings has the properties that are desirable from the point of view of nonlinear elasticity.

In the case of bounded distortion in the complex plane, one has the associated Beltrami equation

$$\bar{\partial} f(z) = \mu(z) \partial f(z),$$

where one assumes that $\|\mu\|_{L^\infty} < 1$. Each mapping of bounded distortion is a solution to this equation and each such an equation with $\|\mu\|_{L^\infty} < 1$ has a homeomorphic solution of bounded distortion. See Sect. 7.3 for more details. It is possible to show the existence of solutions under weaker assumptions (see [4] for exact statements and proofs), like for those compactly supported μ with $\exp(\frac{p}{1-|\mu(z)|})$ integrable for some p; this corresponds to the class of mappings whose distortion is not necessarily bounded but $\exp(\lambda K_f(x)) \in L^1$ for some $\lambda > 0$. Actually O. Lehto proved existence theorems for certain Beltrami coefficients that correspond to suitably integrable distortions K_f already in 1976. We will briefly touch this issue in Sect. 7.3 and we recommend the excellent monograph [4] for the interested reader.

Going back to the question posed by Ball, the following natural issue arises. Given a Sobolev homeomorphism f when do we have that also f^{-1} is a Sobolev mapping or more generally $f^{-1} \in W^{1,p}_{\text{loc}}$? For simplicity, we call a mapping f such that both $f \in W_{\text{loc}}^{1,1}$ and $f^{-1} \in W_{\text{loc}}^{1,1}$ a bi-Sobolev mapping . We will show below that, for a planar Sobolev homeomorphism $f : \Omega \to \mathbf{R}^2$ we have that

$$\int_\Omega K_f(x)\, dx = \int_{f(\Omega)} |Df^{-1}(y)|^2\, dy \,.$$

This statement means that the finiteness of the left-hand side guarantees that $f^{-1} \in W^{1,2}(f(\Omega), \mathbf{R}^2)$. Consequently, the planar minimization problem for $\int K_f$ corresponds to the minimization problem for the harmonic energy of f^{-1}. We will discuss this important issue at the end of Sect. 1.1.

We begin with a discussion on bi-Sobolev mappings that results in the definition of our main concept, a mapping of finite distortion, in the planar, homeomorphic setting. After this, in Sect. 1.2, we recall the definition and basic properties of mappings of bounded distortion. These properties are relevant for the discussion on the deformations in the fourth paragraph of this introduction. We then give the definition of a mapping of finite distortion in the general case. In the following chapters we establish optimal conditions for a mapping of finite distortion to enjoy analogs of the basic properties of mappings of bounded distortion. Especially, we answer J.M.Ball's question described in paragraph two of this introduction (see Sect. 7.2) in the affirmative.

At the end of each chapter and of some of the sections we offer a Remark where we recall the sources of the involved ideas and we also point out the reader's attention to some additional references and results.

1.1 Planar Bi-Sobolev Mappings

Let us first look at the single variable case in order to gain intuition on necessary and sufficient conditions for a homeomorphism to be a bi-Sobolev mapping.

The one dimensional case has a simple solution. Indeed, let $u : \mathbf{R} \to \mathbf{R}$ be an increasing homeomorphism that belongs to $W_{\text{loc}}^{1,1}(\mathbf{R})$; recall that a function v belongs to the Sobolev space $W_{\text{loc}}^{1,1}(\mathbf{R})$ if and only if the restriction of v to each compact interval is absolutely continuous. First of all, if $u^{-1} \in W_{\text{loc}}^{1,1}(\mathbf{R})$, then u^{-1} is absolutely continuous on each compact interval, and hence $|u^{-1}(E)| = 0$ whenever $|E| = 0$. On the other hand,

$$|u(A)| = \int_A u'(x)dx \tag{1.2}$$

for each measurable set A by the (local) absolute continuity of u and the assumption that u is both injective and increasing. Using these observations for the sets $A = \{x : u'(x) = 0\}$ and $E = u(A)$ we thus conclude that necessarily $u'(x) > 0$ almost everywhere. Conversely, suppose that $u \in W_{\text{loc}}^{1,1}(\mathbf{R})$ is an increasing homeomorphism with $u(\mathbf{R}) = \mathbf{R}$ and $u'(x) > 0$ almost everywhere. Then, given a compact interval I, clearly u^{-1} is of bounded variation on I (since u^{-1} is continuous and increasing). Moreover, (1.2) together with the assumption that $u'(x) > 0$ almost everywhere guarantee that u^{-1} maps sets of Lebesgue measure zero to sets of Lebesgue measure zero. Hence u^{-1} is continuous, has bounded variation and maps sets of measure zero to sets of measure zero. It follows that u^{-1} is absolutely continuous on I and thus $u^{-1} \in W_{\text{loc}}^{1,1}(\mathbf{R})$.

Let us move to dimension two. Let $\Omega \subset \mathbf{R}^2$ be a domain and consider a homeomorphism $f : \Omega \to \mathbf{R}^2$ that belongs to $W_{\text{loc}}^{1,1}(\Omega, \mathbf{R}^2)$. This is the usual Sobolev class consisting of all mappings $f : \Omega \to \mathbf{R}^2$ for which both component functions belong to $W_{\text{loc}}^{1,1}(\Omega)$. For the sake of completeness, the Appendix below contains a brief introduction to Sobolev spaces. When do we also have that $f^{-1} \in W_{\text{loc}}^{1,1}(f(\Omega), \mathbf{R}^2)$?

This is not always the case as shown by the following example.

Example 1.1. There is a homeomorphism $f : \mathbf{R}^2 \to \mathbf{R}^2$ such that f is Lipschitz, but $f^{-1} \notin W_{\text{loc}}^{1,1}(\mathbf{R}^2, \mathbf{R}^2)$.

Proof. Indeed, let u be the usual Cantor ternary function (see e.g. [13]) on the interval $(0, 1)$. Then u is continuous, non-decreasing, constant on each complementary interval of the ternary Cantor set and fails to be absolutely continuous. Let now $v(x) = x + u(x)$ on $(0, 1)$ and extend v to negative reals as $v(x) = x$ and to $x \geq 1$ as $v(x) = x + 1$. Then also v fails to be absolutely continuous but v^{-1} is Lipschitz. The mapping g defined simply by $g([x_1, x_2]) = [v(x_1), x_2]$ is clearly a homeomorphism, but it is not absolutely continuous on almost all lines parallel to coordinate axes as v is not absolutely continuous. It follows that g does not satisfy the ACL-condition and hence $g \notin W_{\text{loc}}^{1,1}(\mathbf{R}^2, \mathbf{R}^2)$, see Theorem A.15. It is easy to check that $f = g^{-1}$ is Lipschitz continuous and thus f has the desired properties.
□

Notice that, in the above construction, J_f vanishes in a set of positive area. Based on this and the above discussion on the single variable setting, one could expect that, in the orientation preserving case, the answer should be "if and only if $J_f(x) := \det Df(x) > 0$ almost everywhere." This turns out not to be the correct answer.

Indeed, it is not hard to construct a homeomorphism $f : \mathbf{R}^2 \to \mathbf{R}^2$ that belongs to $W_{\text{loc}}^{1,1}(\mathbf{R}^2, \mathbf{R}^2)$ so that also $f^{-1} \in W_{\text{loc}}^{1,1}(\mathbf{R}^2, \mathbf{R}^2)$ but $J_f(x) = 0$ in a set of positive area. This can be done by mapping a product Cantor-set E of positive area onto a product Cantor-set of area zero via a suitable Lipschitz homeomorphism $f : \mathbf{R}^2 \to \mathbf{R}^2$, see the proof of Theorem 4.15 below for the details. In this construction, one maps squerical frames surrounding E onto substantially smaller squerical frames surrounding the second product Cantor-set in a canonical manner set and then $Df(x) = 0$ (the zero matrix) almost everywhere in the above Cantor-set E and hence almost everywhere in the zero set of the Jacobian of f. We will show below that this phenomenon is both necessary and sufficient for $f^{-1} \in W_{\text{loc}}^{1,1}(f(\Omega), \mathbf{R}^2)$. Towards this end, we begin with a simple oscillation estimate.

Lemma 1.2. *Let* $B(y, 2r) \subset\subset f(\Omega)$ *and suppose that* $f \in W^{1,1}_{\mathrm{loc}}(\Omega, \mathbf{R}^2)$ *is a homeomorphism. Then*

$$r \operatorname{diam} f^{-1}(B(y, r)) \leq \int_{f^{-1}(B(y, 2r))} |Df|. \tag{1.3}$$

Proof. Set $d = \operatorname{diam} f^{-1}(B(y, r))$ and pick $a, b \in \overline{f^{-1}(B(y, r))}$ such that $|a - b| = d$. Without loss of generality, we will suppose that $a = [0, 0]$ and $b = [d, 0]$. For $t \in [0, d]$ we denote

$$L_t = \{ s \in \mathbf{R} : [t, s] \in f^{-1}(B(y, 2r)) \}.$$

Since f is absolutely continuous on almost every line parallel to the y-axis and $\operatorname{diam} f(\{t\} \times L_t) \geq r$, we obtain

$$\int_{L_t} |Df(t, s)| ds \geq r$$

for almost every $t \in [0, d]$. By integrating this inequality over $[0, d]$ we obtain (1.3). $\qquad\square$

Lemma 1.3. *Let* $\Omega \subset \mathbf{R}^2$ *be a domain and let* $f : \Omega \to f(\Omega) \subset \mathbf{R}^2$ *be a homeomorphism. Suppose that both* $f \in W^{1,1}_{\mathrm{loc}}(\Omega, \mathbf{R}^2)$ *and* $f^{-1} \in W^{1,1}_{\mathrm{loc}}(f(\Omega), \mathbf{R}^2)$. *Then* $Df^{-1}(x) = 0$ *almost everywhere in the set* $\{x : J_{f^{-1}}(x) = 0\}$.

Proof. Suppose that we can find a measurable set M and an open set A such that $M \subset A \subset\subset f(\Omega)$, $|A| < \infty$, $|M| > 0$, and

$$\text{for every } x \in M \text{ we have } |Df^{-1}(x)| > 0 \text{ and } J_{f^{-1}}(x) = 0. \tag{1.4}$$

Then there exists $k \in \mathbf{Z}$ such that for

$$\tilde{M} = \{x \in M : 2^k < |Df^{-1}(x)| \leq 2^{k+1}\} \text{ we have } |\tilde{M}| > 0. \tag{1.5}$$

In view of Lemma A.28, we may moreover assume that f^{-1} is differentiable at every point of $\tilde{M}$. Let $\eta > 0$. From (1.4) and (1.5) we obtain that for every $x \in \tilde{M}$ we can pick a disk $B(x, r(x))$ such that

$$B(x, 2r(x)) \subset A, \; r(x) < 1,$$
$$\operatorname{diam} f^{-1}(B(x, r(x))) > 2^k r(x) \text{ and} \tag{1.6}$$
$$|f^{-1}(B(x, 2r(x)))| < \eta |B(x, 2r(x))|.$$

We use the Vitali covering theorem, Theorem A.1, for the family $\{B(x, 2r(x))\}_{x \in \tilde{M}}$ to obtain disks $B_i := B(x_i, r_i)$ such that

$$\tilde{M} \subset \bigcup_i 10 B_i \text{ and } 2B_i \text{ are pairwise disjoint.} \tag{1.7}$$

Hence (1.7), (1.6) and Lemma 1.2 give us

$$|\tilde{M}| \le 10^2 \sum_i |B_i| \le C \sum_i r_i^2$$

$$\le C 2^{-k} \sum_i r_i \operatorname{diam}(f^{-1}(B(x_i, r_i))) \tag{1.8}$$

$$\le C(k) \sum_i \int_{f^{-1}(2B_i)} |Df| = C(k) \int_{\bigcup_i f^{-1}(2B_i)} |Df|.$$

By (1.6) we conclude that

$$\left| \bigcup_i f^{-1}(2B_i) \right| \le \eta \sum_i |2B_i| \le \eta |A| \overset{\eta \to 0+}{\to} 0.$$

Using this fact, (1.8) and the absolute continuity of the integral of $|Df| \in L^1_{\text{loc}}(\Omega)$ we obtain a contradiction with $|\tilde{M}| > 0$. $\qquad\square$

Lemma 1.4. *Let $\Omega \subset \mathbf{R}^2$ be a domain. Suppose that $f : \Omega \to f(\Omega) \subset \mathbf{R}^2$ is a homeomorphism that also belongs to $W^{1,1}_{\text{loc}}(\Omega, \mathbf{R}^2)$. If $|Df(x)| = 0$ almost everywhere in the set $\{x : J_f(x) = 0\}$, then also f^{-1} belongs to $W^{1,1}_{\text{loc}}(f(\Omega), \mathbf{R}^2)$.*

Proof. Let $A \subset\subset f(\Omega)$ be a fixed domain. First we will construct approximations to f^{-1}. We fix $0 < \varepsilon < \frac{1}{4} \operatorname{dist}(A, \partial f(\Omega))$ and we denote the standard ε-grid in $\mathbf{R}^2$ by $G_\varepsilon = (\varepsilon \mathbf{Z}) \times (\varepsilon \mathbf{Z})$. Pick a partition of unity $\{\phi_z\}_{z \in G_\varepsilon}$ such that

each $\phi_z : \mathbf{R}^2 \to \mathbf{R}$ is continuously differentiable;

$$\operatorname{spt} \phi_z \subset B(z, 2\varepsilon) \text{ and } |\nabla \phi_z| \le \frac{C}{\varepsilon}; \tag{1.9}$$

$$\sum_{z \in G_\varepsilon} \phi_z(y) = 1 \text{ for every } y \in \mathbf{R}^2.$$

Now we set

$$g_\varepsilon(y) = \sum_{z \in G_\varepsilon} \phi_z(y) f^{-1}(z) \text{ for every } y \in A.$$

The supports of ϕ_z have bounded overlap and hence this approximation to f^{-1} clearly satisfies $g_\varepsilon \in C^1(A, \mathbf{R}^2)$. Next we show that

$$|Dg_\varepsilon(y)| \le \frac{C}{\varepsilon} \operatorname{diam} f^{-1}(B(y, 2\varepsilon)).$$

Indeed, for a fixed $y \in A$, choose z_0 so that $y \in B(z_0, 2\varepsilon)$. Then

$$Dg_\varepsilon(y) = D \sum_{z \in G_\varepsilon} \phi_z(y)(f^{-1}(z) - f^{-1}(z_0)),$$

by (1.9), and the asserted estimate follows. Together with Lemma 1.2 this implies that for every $y \in A$ we have

$$|Dg_\varepsilon(y)| \le \frac{C}{\varepsilon^2} \int_{f^{-1}(B(y, 4\varepsilon))} |Df(x)| dx. \tag{1.10}$$

Denote $\tilde{G} = \{x \in \Omega : f$ is differentiable at x and $J_f(x) > 0\}$. By Lemma A.28 we know that f is differentiable a.e. and hence we may use our assumptions to conclude that $Df(x) = 0$ a.e. in $\Omega \setminus \tilde{G}$. Pick a Borel set $G \subset \tilde{G}$ such that $|G| = |\tilde{G}|$. From (1.10) and the Area Formula, Corollary A.36 (a), we now have

$$\begin{aligned}
|Dg_\varepsilon(y)| &\le \frac{C}{\varepsilon^2} \int_{f^{-1}(B(y, 4\varepsilon)) \cap G} |Df(x)| dx \\
&\le \frac{C}{\varepsilon^2} \int_{B(y, 4\varepsilon) \cap f(G)} \frac{|(Df)(f^{-1}(z))|}{J_f(f^{-1}(z))} dz.
\end{aligned} \tag{1.11}$$

We claim that

$$F(z) := \frac{|(Df)(f^{-1}(z))|}{J_f(f^{-1}(z))} \chi_{f(G)}(z) \in L^1(A). \tag{1.12}$$

Note that $f(G)$ is a Borel set (as a preimage of a Borel set under the continuous map f^{-1}) and hence F is measurable. For every $z \in f(G)$ we know that f is differentiable at $f^{-1}(z)$ and that $J_f(f^{-1}(z)) > 0$. Therefore f^{-1} is differentiable at z and $J_{f^{-1}}(z) = 1/J_f(f^{-1}(z))$. It follows from the Area Formula, Corollary A.36 (a), for $g = f^{-1}$, that

$$\int_A F(z) dz = \int_{A \cap f(G)} |(Df)(f^{-1}(z))| J_{f^{-1}}(z) dz \le \int_{f^{-1}(A) \cap G} |Df| < \infty.$$

Since $\fint_{B(y, 4\varepsilon)} F \to F(y)$ in $L^1(A)$ as $\varepsilon \to 0$, there is a subsequence $\varepsilon_j \to 0$ such that $\fint_{B(y, 4\varepsilon_j)} F$ has a majorant $H \in L^1(A)$. From this, (1.11) and characterization of weak compactness in L^1, Lemma A.3, we obtain that there is a subsequence $\varepsilon_i \to 0$ and $g \in L^1(A, \mathbf{R}^2)$ such that $Dg_{\varepsilon_i} \to g$ weakly in $L^1(A)$. Clearly

$$\int_A Dg_{\varepsilon_i}(y) \varphi(y) dy = -\int_A g_{\varepsilon_i}(y) D\varphi(y) dy$$

for every test function $\varphi \in C_c^\infty(A, \mathbf{R}^2)$. Since $g_\varepsilon \to f^{-1}$ locally uniformly as $\varepsilon \to 0$, we obtain, after passing to a limit, that

$$\int_A g(y)\varphi(y)dy = -\int_A f^{-1}(y)D\varphi(y)dy$$

which means that g is a weak gradient of f^{-1} in A and therefore $f^{-1} \in W^{1,1}(A, \mathbf{R}^2)$. □

Let us formulate the above necessary and sufficient condition for the Sobolev regularity of f^{-1} as a definition. First of all notice that, assuming that $J_f(x) \geq 0$ almost everywhere, the requirement that $Df(x) = 0$ almost everywhere in the set $\{x : J_f(x) = 0\}$ is equivalent to requiring the existence of a function $K(x) \geq 1$ so that

$$|Df(x)|^2 \leq K(x)J_f(x)$$

almost everywhere. Here one necessarily has $K(x) \geq 1$ by Hadamard's inequality for matrices. Also the almost everywhere differentiability of a homeomorphism $f \in W_{\text{loc}}^{1,1}(\mathbf{R}^2, \mathbf{R}^2)$ guarantees the local integrability of J_f; see Corollary A.36 (a).

Definition 1.5. We say that a homeomorphism $f : \Omega \to f(\Omega) \subset \mathbf{R}^2$ on an open set $\Omega \subset \mathbf{R}^2$ has finite distortion if $f \in W_{\text{loc}}^{1,1}(\Omega, \mathbf{R}^2)$ and there is a function $K : \Omega \to [1, \infty]$ with $K(x) < \infty$ almost everywhere such that

$$|Df(x)|^2 \leq K(x)J_f(x) \qquad \text{for almost all } x \in \Omega.$$

For mappings of finite distortion we can define the optimal distortion function as

$$K_f(x) := \begin{cases} \frac{|Df(x)|^2}{J_f(x)} & \text{for all } x \in \{J_f > 0\}, \\ 1 & \text{for all } x \in \{J_f = 0\}. \end{cases}$$

Using this definition, we can formulate the outcome of Lemmas 1.3 and 1.4 as the following result.

Theorem 1.6. *Let $\Omega \subset \mathbf{R}^2$ be a domain and let $f : \Omega \to f(\Omega) \subset \mathbf{R}^2$ be a homeomorphism with $f \in W_{\text{loc}}^{1,1}(\Omega, \mathbf{R}^2)$ and assume that $J_f(x) \geq 0$ almost everywhere. Then the following conditions are equivalent:*

(i) $f^{-1} \in W_{\text{loc}}^{1,1}(f(\Omega), \mathbf{R}^2)$,

(ii) f *has finite distortion,*

(iii) f^{-1} *has finite distortion.*

Let us recall the question from our introduction regarding the existence and regularity of an inverse of a Sobolev mapping. In our planar setting, the regularity question asks if the inverse of a homeomorphism $f : \Omega \rightarrow f(\Omega)$ that belongs to $W^{1,2}(\Omega, \mathbf{R}^2)$ also belongs to $W^{1,2}(f(\Omega), \mathbf{R}^2)$ under the assumptions that $J_f(x) > 0$ almost everywhere and

$$\int_\Omega |(Df(x))^{-1}|^2 J_f(x)\, dx < \infty.$$

Since $J_f(x) > 0$, our homeomorphism has finite distortion and the above integrability condition guarantees that $K_f \in L^1(\Omega)$. The following theorem completely solves this modified problem.

Theorem 1.7. *Let $\Omega \subset \mathbf{R}^2$ be a domain and let $f : \Omega \rightarrow f(\Omega) \subset \mathbf{R}^2$ be a homeomorphism with $f \in W_{loc}^{1,1}(\Omega, \mathbf{R}^2)$ and assume that $J_f(x) \geq 0$ almost everywhere. Then the following conditions are equivalent:*

(i) $J_f(x) > 0$ almost everywhere and $\int_\Omega |(Df(x))^{-1}|^2 J_f(x)\, dx < \infty$,

(ii) f has finite distortion and $K_f \in L^1(\Omega)$,

(iii) $f^{-1} \in W_{loc}^{1,2}(f(\Omega), \mathbf{R}^2)$ and $|Df^{-1}| \in L^2(f(\Omega))$.

For the proof of this theorem we need the following result from [114] (see also [89]) which is of independent interest.

Theorem 1.8. *Let $\Omega \subset \mathbf{R}^2$ be a domain and let $f \in W_{loc}^{1,2}(\Omega, \mathbf{R}^2)$ be a homeomorphism. Then $|f(E)| = 0$ for every set $E \subset \Omega$ with $|E| = 0$.*

Proof. Let $x \in \Omega$ and $B(x, r) \subset\subset \Omega$. Since f is a homeomorphism we get

$$\mathrm{diam}(f(B(x, r))) = \mathrm{diam}(f(S^1(x, r))) .$$

Each Sobolev mapping (or function) f is absolutely continuous on almost all lines parallel to coordinate axes, Theorem A.15. By a change of variables it follows that f is absolutely continuous on almost all circles. Hence for a.e. $r > 0$ we get by the fundamental theorem of calculus and Hölder's inequality that

$$\mathrm{diam}(f(S^1(x, r))) \leq \int_{S^1(x,r)} |Df| \leq \left(\int_{S^1(x,r)} |Df|^2 \right)^{\frac{1}{2}} \left(\int_{S^1(x,r)} 1 \right)^{\frac{1}{2}} .$$

It follows that

$$\mathrm{diam}^2(f(B(x, r))) = \mathrm{diam}^2(f(S^1(x, r))) \leq Cr \int_{S^1(x,r)} |Df|^2 . \tag{1.13}$$

Let $E \subset \Omega$ satisfy $|E| = 0$. We cover E by disks and we use the previous estimate on each disk in question. We distinguish if $\omega(r) = \int_{B(x,r)} |Df|^2$ decreases too fast or not. Formally, we set

$$E_1 = \left\{ x \in E : \text{essliminf}_{r \to 0+} \frac{r \int_{S^1(x,r)} |Df|^2}{\int_{B(x,r)} |Df|^2} \leq 16 \right\}.$$

and $E_0 = E \setminus E_1$. For every $x \in E_0$ there is $\delta = \delta_x > 0$ such that for a.e. $r \in (0, \delta)$ we have

$$\int_{B(x,r)} |Df|^2 \leq \frac{r}{16} \int_{S^1(x,r)} |Df|^2.$$

Fix $x \in E_0$ and let $\rho \in (0, \delta/2]$. Integrating the previous inequality over the interval $[\rho, 2\rho]$ we obtain

$$\rho \int_{B(x,\rho)} |Df|^2 \leq \frac{\rho}{8} \int_{B(x,2\rho)} |Df|^2.$$

Write $\omega(\rho) = \int_{B(x,\rho)} |Df|^2$. By iteration of the inequality $\omega(\rho) \leq \frac{1}{8}\omega(2\rho)$ we obtain

$$\omega(\rho) = \int_{B(x,\rho)} |Df|^2 \leq \rho^2 \tag{1.14}$$

for small enough $\rho > 0$. Indeed, fix $\rho < \delta/2$ and pick $m \in \mathbf{N}$ such that $\delta/2^m \leq \rho \leq \delta/2^{m-1}$. Now

$$\omega(\rho) \leq \omega\left(\frac{\delta}{2^{m-1}}\right) \leq \frac{1}{8^{m-1}}\omega(\delta) = \frac{8}{8^3}\frac{\delta^3}{8^m}\omega(\delta) \leq \frac{8}{8^3}\omega(\delta)\rho^3$$

and (1.14) follows if we choose ρ sufficiently small. For a fixed $x \in E_0$ we may use (1.14) and the Fubini theorem to find $r \in [\rho/2, \rho]$ such that

$$\int_{S^1(x,r)} |Df|^2 \leq 8r.$$

Choose an open set G such that $E \subset G \subset \Omega$. For every $x \in E$ we may use the definition of E_1 and our previous observation to find $r > 0$ such that $B(x, r) \subset G$ and

$$\int_{S^1(x,r)} |Df|^2 \leq \begin{cases} \frac{16}{r} \int_{B(x,r)} |Df|^2, & x \in E_1, \\ 8r, & x \in E_0. \end{cases}$$

By (1.13) we obtain

$$\text{diam}^2(f(B(x,r))) \leq Cr \int_{S^1(x,r)} |Df|^2 \leq C \int_{B(x,r)} (1 + |Df|^2).$$

Using Besicovitch covering theorem, Theorem A.2, to this collection of disks, we find disks $B_i = B(x_i, r_i)$ that cover E with uniformly bounded overlap. Now

$$|f(E)| \le \sum_i |f(B_i)| \le C \sum_i \mathrm{diam}^2(f(B_i)) \le C \sum_i \int_{B_i} (1 + |Df|^2)$$

$$\le C \int_G (1 + |Df|^2).$$

Letting $|G| \to 0$ we obtain that $|f(E)| = 0$ by the absolute continuity of the integral.   □

Proof (of Theorem 1.7). $(ii) \Rightarrow (iii)$: From Theorem 1.6 we know that $f^{-1} \in W^{1,1}_{loc}$ is a mapping of finite distortion and it remains to show that $|Df^{-1}| \in L^2_{loc}$. Since f^{-1} is a mapping of finite distortion and differentiable a.e. by Lemma A.28, we obtain that

$$\int_{f(\Omega)} |Df^{-1}(y)|^2 dy = \int_A |Df^{-1}(y)|^2 dy$$

where A is a Borel subset of the set

$$G := \{ y \in f(\Omega) : \ f^{-1} \text{ is differentiable at } y \text{ and } J_{f^{-1}}(y) > 0 \}$$

such that $|A| = |G|$. It is easy to see that f is differentiable at every point $x \in f^{-1}(A)$ and we have $Df^{-1}(f(x)) = (Df(x))^{-1}$ and $J_f(x) = (J_{f^{-1}}(f(x)))^{-1}$ (see Lemma A.29). Applying these facts, the Area formula for f^{-1}, Corollary A.36 (c), and $E \, \mathrm{adj} \, E = I \det E$ we arrive at

$$\int_{f(\Omega)} |Df^{-1}(y)|^2 dy = \int_A \frac{|Df^{-1}(y)|^2}{J_{f^{-1}}(y)} J_{f^{-1}}(y) dy = \int_{f^{-1}(A)} \frac{|Df^{-1}(f(x))|^2}{J_{f^{-1}}(f(x))} dx$$

$$= \int_{f^{-1}(A)} |(Df(x))^{-1}|^2 J_f(x) dx = \int_{f^{-1}(A)} \frac{|\mathrm{adj} \, Df(x)|^2}{J_f(x)} dx$$

$$= \int_{f^{-1}(A)} \frac{|Df(x)|^2}{J_f(x)} dx \le \int_{\Omega} K_f(x) \, dx < \infty . \tag{1.15}$$

$(iii) \Rightarrow (i)$: Suppose for contrary that $|\{J_f = 0\}| > 0$. Then we may use a.e. differentiability of f, Lemma A.28, to find a Borel subset $\tilde{A} \subset \{J_f = 0\}$ of full measure such that f is differentiable at every point of $\tilde{A}$. By the Area formula, Corollary A.36 (c), we get

$$0 = \int_{\tilde{A}} J_f(x) \, dx = \int_{f(\tilde{A})} 1 \, dy$$

and hence $|f(\tilde{A})| = 0$. We may apply Theorem 1.8 to f^{-1} to obtain that $|f^{-1}(E)| = 0$ for every set $E \subset f(\Omega)$ such that $|E| = 0$. This observation for $E = f(\tilde{A})$ gives us the contradiction and hence $J_f > 0$ a.e.

We can find a Borel set $\tilde{A}$ which is a subset of

$$\tilde{G} := \{x \in \Omega : \ f \text{ is differentiable at } x \text{ and } J_f(x) > 0\}$$

with $|\tilde{A}| = |\tilde{G}|$. By differentiability of f a.e. and the Area formula, Corollary A.36 (c), we obtain analogously to (1.15) that

$$\int_\Omega |(Df(x))^{-1}|^2 J_f(x)\, dx = \int_{\tilde{A}} |(Df(x))^{-1}|^2 J_f(x)\, dx$$

$$= \int_{\tilde{A}} |Df^{-1}(f(x))|^2 J_f(x)\, dx$$

$$= \int_{f(\tilde{A})} |Df^{-1}(y)|^2\, dy < \infty\,.$$

$(i) \Rightarrow (ii)$: Since $J_f > 0$ a.e. it is immediate that f has finite distortion. Since f is differentiable a.e. by Lemma A.28, we may integrate over the set

$$\tilde{G} := \{x \in \Omega : \ f \text{ is differentiable at } x \text{ and } J_f(x) > 0\}$$

and analogously to (1.15) we obtain

$$\int_\Omega K_f(x)\, dx = \int_{\tilde{G}} K_f(x)\, dx = \int_{\tilde{G}} \frac{|\operatorname{adj} Df(x)|^2}{J_f(x)} dx$$

$$= \int_{\tilde{G}} |(Df(x))^{-1}|^2 J_f(x)\, dx < \infty\,. \qquad \square$$

Corollary 1.9. *Let $f \in W_{\text{loc}}^{1,1}(\Omega, f(\Omega))$ be a homeomorphism in the plane with finite distortion. Then*

$$\int_\Omega K_f(x)\, dx = \int_{f(\Omega)} |Df^{-1}(y)|^2\, dy\,.$$

Proof. We may clearly assume that one and hence, by Theorem 1.7, both of these integrals are finite. By the statement of Theorem 1.7 and the first part of its proof it is enough to show that we have an identity in (1.15). This amounts to the claim that

$$\int_{\Omega \setminus f^{-1}(A)} K_f(x)\, dx = 0\,.$$

To prove this claim it suffices to prove that $|\Omega \setminus f^{-1}(A)| = 0$. By the Area formula, Corollary A.36 (c), we get that

$$|f^{-1}(S)| = 0 \quad \text{for} \quad S := \{y \in f(\Omega) : f^{-1} \text{ is differentiable at } y \text{ and } J_{f^{-1}}(y) = 0\}.$$

Since $|f(\Omega) \setminus (A \cup S)| = 0$ and $f^{-1} \in W^{1,2}_{\text{loc}}$ we can use Theorem 1.8 to conclude

$$|f^{-1}(f(\Omega) \setminus (A \cup S))| = 0 \quad \text{and hence} \quad |f^{-1}(f(\Omega) \setminus A)| = 0. \qquad \square$$

Remark 1.10. The definition of K_f is based on the operator norm of Df. Another natural choice is the Hilbert-Schmidt norm

$$\|A\|^2 = \frac{1}{2} \text{Trace}(A^T A) = \frac{1}{2} \sum_{i=1}^{n} \sum_{j=1}^{n} a_{ij}^2 .$$

The identity in Corollary 1.9 also holds when the operator norm is replaced by the Hilbert-Schmidt norm and K_f is replaced with the associated distortion $\mathbb{K}_f$. Notice that the Hilbert-Schmidt norm is strictly convex. Hence the minimization problem for $\int_\Omega \mathbb{K}_f$ for planar homeomorphisms $f : \Omega \to \Omega'$, that coincide on the boundary with a fixed homeomorphism $g : \overline{\Omega} \to \overline{\Omega'}$ with $\int_\Omega \mathbb{K}_g < \infty$, assuming that Ω is convex, has a unique (modulo conformal changes of variables) minimizer and its inverse is a harmonic mapping. For this see [5, 6] and [52]. Regarding powers of the distortion function, it was shown in [6] that, apart from some trivial cases, the minimizers of the L^p-norm of the distortion function never exist when $p < 1$.

Open problem 1. Let Ω be a convex planar domain and $g : \overline{\Omega} \to \overline{\Omega'}$ be a homeomorphism of finite distortion with $\int_\Omega \mathbb{K}_g^p < \infty$, where $p > 1$. Does the minimization problem of $\int_\Omega \mathbb{K}_f^p$ for homeomorphisms $f : \Omega \to \Omega'$ that coincide on the boundary with g have a diffeomorphic solution?

The following open problems deal with minimization without given boundary values. In nonlinear elasticity, this corresponds to traction free problems.

Open problem 2. Let Ω, Ω' be bounded doubly connected planar domains. Then the minimization problem for $\int_\Omega \mathbb{K}_f$ for homeomorphisms $f : \Omega \to \Omega'$ does not necessarily have a homeomorphic solution. There is a homeomorphic minimizer if $\text{Mod} \, \Omega \geq \text{Mod} \, \Omega'$, where $\text{Mod} \, G$ refers to the conformal modulus. Prove that the existence of a homeomorphic minimizer implies that necessarily $\text{Mod} \, \Omega' \leq \log \cosh \text{Mod} \, \Omega$. See [5, 64, 68] for the definition of the conformal modulus and more details.

Open problem 3. Let Ω, Ω' be multiply connected planar domains such that there exists a homeomorphism $g : \Omega \to \Omega'$ of finite distortion with $\mathbb{K}_g \in L^p(\Omega)$, where $p > 1$. Prove that there is a homeomorphism $f : \Omega \to \Omega'$ that minimizes $\int_\Omega \mathbb{K}_f^p$. For related results see [93].

1.2 Mappings of Bounded and Finite Distortion

Let Ω be an open connected set in $\mathbf{R}^n$ for some $n \geq 2$. Then a mapping $f : \Omega \to \mathbf{R}^n$ is called quasiregular or a mapping of bounded distortion if $f \in W^{1,n}_{\text{loc}}(\Omega, \mathbf{R}^n)$ and there is a constant $K \geq 1$ so that

$$|Df(x)|^n \leq KJ_f(x)$$

almost everywhere in Ω. It is then customary to say that f is K-quasiregular. This class of mappings was introduced by Reshetnyak in 1967 [115]. We recommend the monographs [75, 116, 117] for an interested reader.

Let us list some of the basic properties of quasiregular mappings. Let $f : \Omega \to \mathbf{R}^n$ be a K-quasiregular mapping.

- The mapping f has a continuous representative; actually a $1/K$-Hölder continuous one.
- This continuous representative $\hat{f}$ is either constant or both open and discrete: images of open sets are open and the preimage of no point can accumulate in Ω. In the latter case, $\hat{f}$ is locally bounded-to-one, and if Ω is bounded and $\hat{f}$ is injective close to the boundary of Ω, then $\hat{f}$ is a homeomorphism.
- The mapping $\hat{f}$ maps sets of measure zero to sets of measure zero. In the case of a non-constant $\hat{f}$, preimages of sets of measure zero are sets of measure zero and thus the Jacobian determinant is necessarily strictly positive almost everywhere.
- Homeomorphic quasiregular (i.e. quasiconformal) mappings form a group with respect to composition: f^{-1} is K^{n-1}-quasiregular if f is K-quasiregular, and $f_1 \circ f_2$ is $K_1 K_2$-quasiregular whenever defined, if f_1 is K_1-quasiregular and f_2 is K_2-quasiregular.
- Regarding regularity, there is $p = p(n, K) > n$ so that each K-quasiregular mapping f belongs to $W^{1,p}_{\text{loc}}(\Omega, \mathbf{R}^n)$ and so that J_f^{n-p} is locally integrable (unless f is constant). For homeomorphic quasiregular mapping this implies that also $J_{f^{-1}}^{-\varepsilon}$ is locally integrable.

Definition 1.11. We say that a mapping $f : \Omega \to \mathbf{R}^n$ on an open connected set $\Omega \subset \mathbf{R}^n$ has finite distortion if $f \in W^{1,1}_{\text{loc}}(\Omega, \mathbf{R}^n)$, $J_f \in L^1_{\text{loc}}(\Omega)$ and there is a function $K : \Omega \to [1, \infty]$ with $K(x) < \infty$ almost everywhere such that

$$|Df(x)|^n \leq K(x)J_f(x) \qquad \text{for almost all } x \in \Omega. \tag{1.16}$$

For mappings of finite distortion we can define the optimal distortion function as

$$K_f(x) := \begin{cases} \frac{|Df(x)|^n}{J_f(x)} & \text{for all } x \in \{J_f > 0\}, \\ 1 & \text{for all } x \in \{J_f = 0\}. \end{cases}$$

In the next chapters we relax the assumption $K \in L^\infty$ and we prove that mappings of finite distortion have properties similar to those of mappings of bounded distortion. We usually have two kinds of positive results. We assume that f is in the nice Sobolev space $W^{1,n}$ and then we require some mild assumptions on the distortion like integrability or only finiteness almost everywhere. Alternatively, we assume only that $f \in W^{1,1}$ but then we usually need much stronger assumptions on the distortion, like $\exp(\lambda K_f) \in L^1$ for some $\lambda > 0$.

Chapter 2
Continuity

Abstract The main aim of this chapter is to establish sufficient conditions for continuity for mappings of finite distortion. This corresponds to the property that material does not break during the deformation f of a body in $\mathbf{R}^n$ and that no cavities are created in the interior of the body Ω under the deformation.

2.1 Counterexamples and Idea of the Proof

First we will study natural counterexamples in order to understand the assumptions necessary for continuity. Then we will briefly sketch the main idea of the proof.

Lemma 2.1. *Let $\rho : (0, \infty) \rightarrow (0, \infty)$ be a strictly monotone function with $\rho \in C^1((0, \infty))$. Then, for the mapping*

$$f(x) = \frac{x}{|x|} \rho(|x|), \quad x \neq 0,$$

we have for almost every x

$$|Df(x)| = \max\left\{\frac{\rho(|x|)}{|x|}, |\rho'(|x|)|\right\} \text{ and } J_f(x) = \rho'(|x|)\left(\frac{\rho(|x|)}{|x|}\right)^{n-1}.$$

Proof. The idea is that the derivative of our function f at the point x in the direction of $\frac{x}{|x|}$ equals $\rho'(|x|)$ and the derivative in any tangential direction equals $\frac{\rho(|x|)}{|x|}$. The last claim can be easily seen from the fact that the sphere of radius $|x|$ centered at the origin is mapped to a similar sphere of radius $\rho(|x|)$, the derivative along the sphere is the same as the derivative in the direction parallel to the sphere and by symmetry the norm of the derivative is constant on the sphere.

It is not difficult to see that $f \in C^1(\mathbf{R}^n \setminus \{0\})$. By radial symmetry it is enough to show the result only for a single point at each sphere. Let us fix a point $x = [x_1, 0, \ldots, 0]$ such that $x_1 > 0$. By a direct computation

S. Hencl and P. Koskela, *Lectures on Mappings of Finite Distortion*, Lecture Notes in Mathematics 2096, DOI 10.1007/978-3-319-03173-6_2,
© Springer International Publishing Switzerland 2014

$$\frac{\partial f_1}{\partial x_1}(x) = \lim_{t\to 0} \frac{\frac{x_1+t}{|x_1+t|}\rho(x_1+t) - \frac{x_1}{|x_1|}\rho(x_1)}{t} = \lim_{t\to 0} \frac{\rho(x_1+t) - \rho(x_1)}{t}$$

$$= \rho'(x_1) = \rho'(|x|)$$

and for every $i \in \{2,\ldots,n\}$ we have

$$\frac{\partial f_i}{\partial x_i}(x) = \lim_{t\to 0} \frac{\frac{t}{|x_1\mathbf{e}_1+t\mathbf{e}_i|}\rho(|x_1\mathbf{e}_1+t\mathbf{e}_i|) - \frac{0}{|x_1|}\rho(x_1)}{t} = \frac{\rho(x_1)}{|x_1|} = \frac{\rho(|x|)}{|x|},$$

where we used the notation $\mathbf{e}_i = (0,\ldots,0,1,0,\ldots,0)$ for the i-th unit vector. Moreover

$$\frac{\partial f_i}{\partial x_1}(x) = \lim_{t\to 0} \frac{\frac{0}{|x_1+t|}\rho(|x_1+t|) - \frac{0}{|x_1|}\rho(x_1)}{t} = 0$$

and analogously it is easy to see that $\frac{\partial f_i}{\partial x_j}(x) = 0$ for every $i \neq 1$ and $i \neq j$. It remains to consider $i \in \{2,\ldots,n\}$ and to compute

$$\frac{\partial f_1}{\partial x_i}(x) = \lim_{t\to 0} \frac{\frac{x_1}{|x_1\mathbf{e}_1+t\mathbf{e}_i|}\rho(|x_1\mathbf{e}_1+t\mathbf{e}_i|) - \frac{x_1}{|x_1|}\rho(x_1)}{t}$$

$$= \lim_{t\to 0} \frac{\frac{x_1}{|x_1\mathbf{e}_1+t\mathbf{e}_i|} - 1}{t}\rho(|x_1\mathbf{e}_1+t\mathbf{e}_i|) + \lim_{t\to 0} \frac{\rho(|x_1\mathbf{e}_1+t\mathbf{e}_i|) - \rho(x_1)}{t}.$$

$$(2.1)$$

Since $\mathbf{e}_1 \perp \mathbf{e}_i$ it is easy to see that

$$\lim_{t\to 0} \frac{|x_1\mathbf{e}_1+t\mathbf{e}_i| - x_1}{t} = 0$$

which implies that the first limit on the right-hand side of (2.1) is zero. Together with the chain rule it also implies that the second limit is zero and hence $\frac{\partial f_1}{\partial x_i}(x) = 0$. It is now easy to compute the norm of the derivative and the Jacobian determinant. □

Let us focus on the following crucial example of the so-called cavitation phenomenon.

Example 2.2. For $x \in B(0,1) \setminus \{0\}$ let us set

$$f(x) = x + \frac{x}{|x|}$$

(and define $f(0) = 0$). Then f is a mapping of finite distortion such that $f \in W^{1,p}(B(0,1), \mathbf{R}^n)$ for all $p < n$, but f is not continuous at the origin, i.e. no representative of f is continuous at the origin.

Proof. Clearly $f(x) = \frac{x}{|x|}(|x| + 1)$ maps spheres centered at the origin onto similar spheres and it is a diffeomorphism from $B(0, 1) \setminus \{0\}$ onto $B(0, 2) \setminus \overline{B(0, 1)}$. Using Lemma 2.1 it is easy to check that

$$|Df(x)| = 1 + \frac{1}{|x|} \text{ and } J_f(x) = \left(1 + \frac{1}{|x|}\right)^{n-1}.$$

It follows (see Theorem A.15 (1)) that $f \in W^{1,p}$ for all $p < n$ and even that

$$\frac{|Df|^n}{\log^{1+\varepsilon}(e + |Df|)} \in L^1 \text{ for all } \varepsilon > 0.$$

Also, $J_f \in L^1(B(0, 1))$ and, since $J_f > 0$ almost everywhere, it follows that f has finite distortion. On the other hand, we cannot extend f continuously to the origin since each sphere $S^{n-1}(0, \varepsilon)$ is mapped to the sphere $S^{n-1}(0, 1 + \varepsilon)$. □

It is a well-known fact that each function in the Sobolev space $W^{1,p}$ has a continuous representative if $p > n$ but not necessarily for any $p \le n$. From the previous example we know that we cannot hope for a positive result if we only know that a mapping f of finite distortion belongs to $W^{1,p}$ for some $p < n$. The following result that we will prove later in this chapter shows that in the limiting situation $f \in W^{1,n}$ mappings of finite distortion have better properties than general Sobolev mappings.

Theorem 2.3. *Let $\Omega \subset \mathbf{R}^n$ be open and let $f \in W^{1,n}_{loc}(\Omega, \mathbf{R}^n)$ be a mapping of finite distortion. Then f has a continuous representative.*

Moreover, we can relax the regularity assumption on f if we require additional restrictions on the integrability of the distortion function.

Theorem 2.4. *Let $\Omega \subset \mathbf{R}^n$ be open and let $f : \Omega \to \mathbf{R}^n$ be a mapping of finite distortion. Suppose that there is $\lambda > 0$ such that $\exp(\lambda K) \in L^1_{loc}(\Omega)$. Then f has a continuous representative.*

To simplify our notation, from this point on we will say that our mapping of finite distortion is continuous if it has a continuous representative. The following example shows that the integrability assumptions on the distortion function cannot be essentially relaxed for a general $W^{1,1}$-mapping of finite distortion.

Example 2.5. Let $0 < \delta < 1$ and $n \ge 2$. There exists $f \in W^{1,1}(B(0, 1/2), \mathbf{R}^n)$ with finite distortion such that $\exp(\lambda K_f^{1-\delta}) \in L^1(B(0, 1/2))$ for all $\lambda > 0$ but so that f is not continuous at 0.

Proof. Put $\varepsilon := (1 - \frac{\delta}{2})^{-1} - 1$ and for $x \in B(0, 1) \setminus \{0\}$ let us define

$$f(x) := \frac{x}{|x|}\left(1 + \log^{-\varepsilon}\left(\frac{1}{|x|}\right)\right).$$

Then f is a diffeomorphism of $B(0, 1/2) \setminus \{0\}$ onto $B(0, 1 + \log^{-\varepsilon}(2)) \setminus \overline{B(0, 1)}$. Using Lemma 2.1 it is easy to check that for $|x|$ small enough we have

$$|Df(x)| = \frac{1 + \log^{-\varepsilon}\left(\frac{1}{|x|}\right)}{|x|} \text{ and}$$

$$J_f(x) = \varepsilon \frac{\log^{-1-\varepsilon}\left(\frac{1}{|x|}\right)}{|x|} \frac{\left(1 + \log^{-\varepsilon}\left(\frac{1}{|x|}\right)\right)^{n-1}}{|x|^{n-1}}.$$

Hence (see Theorem A.15 (1)) $f \in W^{1,1}$, $J_f \in L^1$, and it easily follows that f is a mapping of finite distortion. Moreover, we have

$$K_f(x) = \varepsilon^{-1} \log^{\frac{1}{1-\delta/2}}\left(\frac{1}{|x|}\right) + \varepsilon^{-1} \log\left(\frac{1}{|x|}\right)$$

when $|x|$ is small and K_f is uniformly bounded in the rest of $B(0, 1/2)$. This easily gives that

$$\exp(\tfrac{\varepsilon}{2}(K_f(x))^{1-\delta/2}) \leq \frac{c}{|x|},$$

which is in $L^1(B(0, 1))$. It now follows that $\exp(\lambda K_f^{1-\delta}) \in L^1(B(0, 1/2))$ for all $\lambda > 0$ and similarly to Example 2.2 we know that f is not continuous at the origin. □

Now we would like to briefly sketch the idea of the proof of Theorem 2.3 and highlight the technical difficulties that we will need to face. Suppose, for simplicity, that $n = 2$ and that f is (an orientation preserving) diffeomorphism outside the origin, and that we want estimate the oscillation of f near the origin. By the fundamental theorem of calculus and Hölder's inequality we obtain

$$\text{diam } f(S^1(0, t)) \leq \int_{S^1(0,t)} |Df| \leq \left(\int_{S^1(0,t)} |Df|^2\right)^{\frac{1}{2}} (2\pi t)^{\frac{1}{2}}. \tag{2.2}$$

It follows that

$$\frac{\text{diam}^2 f(S^1(0, r))}{2\pi} \underset{\sim}{\leq} \frac{1}{\log 2} \int_r^{2r} \frac{\text{diam}^2 f(S^1(0, t))}{2\pi t} \, dt$$

$$\underset{\sim}{\leq} \frac{1}{\log 2} \int_r^{2r} \int_{S^1(0,t)} |Df|^2 \, dt \leq \frac{1}{\log 2} \int_{B(0,2r)} |Df|^2. \tag{2.3}$$

Hence

$$\text{diam } f(B(0, r)) \leq \text{diam } f(S^1(0, r)) \tag{2.4}$$

and by (2.3) the above tends to zero as $r \to 0+$ which implies continuity at the origin.

Of course these inequalities are not valid for general mappings and we need to find an appropriate substitute. It is possible to estimate diam $f(S^{n-1}(0, r))$ by the integral of the derivative, for most of the spheres, for general Sobolev mappings by the Sobolev embedding theorem on spheres and this will give us a suitable analog of (2.2). The most difficult step is to establish an analog of (2.4). To get this estimate we need to show that, under our assumptions, our mapping f is in some sense monotone. For this we will use the nonnegativity of the Jacobian and we will show that the Jacobian coincides with the so-called distributional Jacobian. This will allow us to prove a version of monotonicity called weak monotonicity.

We also need to prove Theorem 2.4. On the first line of (2.3) we have essentially used the fact that $\int \frac{1}{t} = \infty$ but it would also be possible to use $\int \frac{1}{t \log 1/t} = \infty$ and to integrate with respect to different bounds. Then it would be possible to use a general version of Hölder's inequality in (2.2) and we would obtain that it is enough to know that $|Df|^n / \log(e + |Df|) \in L^1$. This integrability condition is actually true under the assumptions of Theorem 2.4 and we will prove it soon.

Definition 2.6. Let $\Omega \subset \mathbf{R}^n$ be an open set, $p \in [1, \infty)$ and $\alpha \in \mathbf{R}$. We say that $H : \Omega \to \mathbf{R}$ belongs to the space $L^p \log^\alpha L(\Omega)$ if

$$\int_\Omega |H(x)|^p \log^\alpha (e + |H(x)|)\, dx < \infty\,.$$

We say that $H \in L^p \log^\alpha L_{\mathrm{loc}}(\Omega)$ if $H \in L^p \log^\alpha L(\tilde{\Omega})$ for all subdomains $\tilde{\Omega} \subset\subset \Omega$.

We define the Sobolev Zygmund space $WL^p \log^\alpha L(\Omega)$ as

$$WL^p \log^\alpha L(\Omega) = \{u \in L^p \log^\alpha L(\Omega) : |Du| \in L^p \log^\alpha L(\Omega)\}$$

where Du denotes the weak derivative as in Definition A.13.

We will need the following special version of Jensen's inequality.

Lemma 2.7. *For $a \geq 1$, $b \geq 0$ and $\lambda > 0$ it holds that*

$$ab \leq \exp(\lambda a) + \frac{2b}{\lambda} \log\left(e + \frac{b}{\lambda}\right).$$

Proof. If $ab \leq \exp(\lambda a)$, then the inequality holds. Thus we may assume that

$$\exp(\lambda a) < ab.$$

In this case, the inequality $x^2 < \exp(x)$ for all $x \geq 0$ implies that

$$\lambda^2 a^2 \leq \exp(\lambda a) < ab.$$

Therefore we get

$$\exp(\lambda a) \le ab < \frac{b^2}{\lambda^2},$$

which implies

$$ab < \frac{2b}{\lambda} \log\left(e + \frac{b}{\lambda}\right). \qquad \square$$

Lemma 2.8. *Let* $\Omega \subset \mathbf{R}^n$ *be open. Let* $f \in W_{\mathrm{loc}}^{1,1}(\Omega, \mathbf{R}^n)$ *have finite distortion and suppose there is* $\lambda > 0$ *such that* $\exp(\lambda K_f) \in L_{\mathrm{loc}}^1(\Omega)$. *Then* $|Df| \in L^n \log^{-1} L_{\mathrm{loc}}(\Omega)$.

Proof. Using the distortion inequality (1.16) and the assumptions $K_f \ge 1$ and $J_f \ge 0$ almost everywhere, we have

$$\frac{|Df(x)|^n}{\log(e + |Df(x)|)} \le \frac{K_f(x)J_f(x)}{\log(e + K_f^{\frac{1}{n}}(x)J_f^{\frac{1}{n}}(x))} \le \frac{K_f(x)J_f(x)}{\log(e + J_f^{\frac{1}{n}}(x))} \le n\frac{K_f(x)J_f(x)}{\log(e + J_f(x))}.$$

Let $U \subset\subset \Omega$. We use Lemma 2.7 for $a := K_f$ and $b := \frac{J_f}{\log(e+J_f)}$ to obtain

$$\int_U \frac{|Df|^n}{\log(e + |Df|)} \le n \int_U \frac{K_f J_f}{\log(e + J_f)}$$

$$\le n \int_U \exp(\lambda K_f)$$

$$+ \frac{2n}{\lambda} \int_U \frac{J_f}{\log(e + J_f)} \log\left(e + \frac{J_f}{\lambda \log(e + J_f)}\right).$$

The first of these two terms is finite by our assumptions. We separate the second integral into integrals over the sets $A_1 := \{x \in U : \lambda \log(e + J_f(x)) \le 1\}$ and $A_2 := U \setminus A_1$. The integrand is bounded on A_1 and on A_2 it is dominated by CJ_f, which is integrable over U. $\qquad \square$

2.2 Distributional Jacobian

Definition 2.9. Let $f \in W^{1,\frac{n^2}{n+1}}(\Omega, \mathbf{R}^n)$. The distributional Jacobian of f is the distribution defined by setting

$$\mathscr{J}_f(\varphi) := -\int_\Omega f_1(x) J(\varphi, f_2, \dots, f_n)(x)\, dx \qquad \text{for all } \varphi \in C_C^\infty(\Omega), \quad (2.5)$$

where $J(\varphi, f_2, \ldots, f_n)$ is the classical Jacobian defined as the determinant of the Jacobi matrix Dg of $g = (\varphi, f_2, \ldots, f_n)$.

It is important that, for sufficiently smooth functions, the distributional Jacobian coincides with the usual Jacobian, i.e. (2.6) below holds.

Proposition 2.10. *For every function $f \in C^2(\Omega, \mathbf{R}^n)$ we have*

$$\int_\Omega \varphi J_f = - \int_\Omega f_1 J(\varphi, f_2, \ldots, f_n) \qquad \text{for all } \varphi \in C_C^\infty(\Omega) . \qquad (2.6)$$

Moreover, the same extends to hold for every $f \in W^{1,n}(\Omega, \mathbf{R}^n)$.

Proof. For simplicity we will sketch the proof only in dimension $n = 2$. The same idea works in arbitrary dimension but the summation formulas are longer.

Let us assume that f is C^2. By integration by parts and the interchangeability of second derivatives we obtain

$$\int_\Omega \left(\frac{\partial f_1}{\partial x_1} \frac{\partial f_2}{\partial x_2} - \frac{\partial f_1}{\partial x_2} \frac{\partial f_2}{\partial x_1} \right) \varphi = - \int_\Omega f_1 \left(\frac{\partial}{\partial x_1} \left(\frac{\partial f_2}{\partial x_2} \varphi \right) - \frac{\partial}{\partial x_2} \left(\frac{\partial f_2}{\partial x_1} \varphi \right) \right)$$

$$= - \int_\Omega f_1 \left(\frac{\partial^2 f_2}{\partial x_1 \partial x_2} \varphi + \frac{\partial f_2}{\partial x_2} \frac{\partial \varphi}{\partial x_1} - \frac{\partial^2 f_2}{\partial x_2 \partial x_1} \varphi - \frac{\partial f_2}{\partial x_1} \frac{\partial \varphi}{\partial x_2} \right)$$

$$= - \int_\Omega f_1 \left(\frac{\partial f_2}{\partial x_2} \frac{\partial \varphi}{\partial x_1} - \frac{\partial f_2}{\partial x_1} \frac{\partial \varphi}{\partial x_2} \right) = - \int_\Omega f_1 J(\varphi, f_2) .$$

Hence (2.6) holds for an arbitrary C^2-function and now let us assume $f \in W^{1,2}$. We find a sequence of functions $f^k \in C^2$ such that $f^k \to f$ in $W^{1,2}_{\text{loc}}$ (see Theorem A.15). We may write the previous equality for each f^k and after passing to the limit we obtain the same formula for f. For example for the first term of the left-hand side we have

$$\int_\Omega \left| \frac{\partial f_1^k}{\partial x_1} \frac{\partial f_2^k}{\partial x_2} \varphi - \frac{\partial f_1}{\partial x_1} \frac{\partial f_2}{\partial x_2} \varphi \right|$$

$$\leq \int_\Omega \left| \frac{\partial f_1^k}{\partial x_1} \frac{\partial f_2^k}{\partial x_2} \varphi - \frac{\partial f_1^k}{\partial x_1} \frac{\partial f_2}{\partial x_2} \varphi \right| + \int_\Omega \left| \frac{\partial f_1^k}{\partial x_1} \frac{\partial f_2}{\partial x_2} \varphi - \frac{\partial f_1}{\partial x_1} \frac{\partial f_2}{\partial x_2} \varphi \right|$$

$$\leq \|\varphi\|_\infty \left(\int_\Omega \left| \frac{\partial f_1^k}{\partial x_1} \right|^2 \right)^{\frac{1}{2}} \left(\int_\Omega \left| \frac{\partial f_2^k}{\partial x_2} - \frac{\partial f_2}{\partial x_2} \right|^2 \right)^{\frac{1}{2}} +$$

$$+ \|\varphi\|_\infty \left(\int_\Omega \left| \frac{\partial f_2}{\partial x_2} \right|^2 \right)^{\frac{1}{2}} \left(\int_\Omega \left| \frac{\partial f_1^k}{\partial x_1} - \frac{\partial f_1}{\partial x_1} \right|^2 \right)^{\frac{1}{2}} \xrightarrow{k \to \infty} 0 . \qquad \square$$

Remark 2.11. (a) The integral in (2.5) is finite for each $f \in W^{1,\frac{n^2}{n+1}}(\Omega, \mathbf{R}^n)$. Indeed, by the Sobolev Embedding Theorem A.18 we know that each $f \in W^{1,\frac{n^2}{n+1}}_{\text{loc}}$ belongs to $L^{p^*}_{\text{loc}}$ for $p^* = \frac{n\frac{n^2}{n+1}}{n-\frac{n^2}{n+1}} = n^2$. Hence we may use Hölder's inequality to obtain

$$\left| \int_\Omega f_1 J(\varphi, f_2, \ldots, f_n) \right| \leq \|f_1\|_{L^{n^2}(\text{spt}\,\varphi)} \||Df|^{n-1} \sup_{x\in\Omega} |\nabla\varphi(x)|\|_{L^{\frac{n^2}{n^2-1}}(\text{spt}\,\varphi)}$$

which is finite since $(n-1)\frac{n^2}{n^2-1} = \frac{n^2}{n+1}$ and $f \in W^{1,\frac{n^2}{n+1}}_{\text{loc}}$.

(b) For $f \in W^{1,p}$, $p < n$, it may well happen that $J_f \neq \mathcal{J}_f$. For example for the mapping

$$f(x) = \frac{x}{|x|} \quad \text{on } B(0,1) \setminus \{0\} \text{ and } f(0) = 0 \text{ we have } \mathcal{J}_f = |B(0,1)|\delta_0$$

where δ_0 denotes the Dirac measure at 0. To show this let, us denote for $k \in \mathbf{N}$

$$f^k(x) = x\left(\frac{1+k}{1+k|x|^2}\right)^{\frac{1}{2}}.$$

These functions are smooth and hence we know from the previous proposition that

$$\int_{B(0,1)} \varphi J_{f^k} = -\int_{B(0,1)} f_1^k J(\varphi, f_2^k, \ldots, f_n^k) \quad \text{for all } \varphi \in C^\infty_C(B(0,1)).$$

It is not difficult to check that $f^k \to f$ in $W^{1,p}$ for all $1 \leq p < n$ and using estimates analogous to part (a) we obtain

$$\lim_{k\to\infty} \int_{B(0,1)} f_1^k J(\varphi, f_2^k, \ldots, f_n^k) = \int_{B(0,1)} f_1 J(\varphi, f_2, \ldots, f_n).$$

To show our conclusion it is now enough to show that

$$\lim_{k\to\infty} \int_{B(0,1)} \varphi J_{f^k} = \int_{B(0,1)} \varphi \, d(|B(0,1)|\delta_0) = |B(0,1)|\varphi(0). \qquad (2.7)$$

Using Lemma 2.1 it is not difficult to show that

$$|J_{f^k}(x)| \leq \frac{C}{|x|^n} \quad \text{and} \quad \lim_{k\to\infty} J_{f^k}(x) = 0 \text{ for a.e. } x \in B(0,1).$$

Therefore

$$|\varphi(x) - \varphi(0)| \cdot |J_{f^k}(x)| \le |x| \sup |\nabla \varphi| \frac{C}{|x|^n} .$$

It is easy to see that $\int_{B(0,1)} J_{f_k} = 1$ for all k and hence we may use the Lebesgue dominated convergence theorem to conclude that

$$\lim_{k \to \infty} \int_{B(0,1)} (\varphi(x) - \varphi(0)) J_{f^k}(x) \, dx = 0 .$$

On the other hand, $\int_{B(0,1)} J_{f_k} = 1$ for all k, and (2.7) and our statement follow.

In the proof of Theorems 2.3 and 2.4 we will essentially need that the distributional Jacobian coincides with the pointwise Jacobian. For Theorem 2.3 we may apply Proposition 2.10 but for Theorem 2.4 we will need the following stronger result.

Theorem 2.12. *Let $\Omega \subset \mathbf{R}^n$ be open. Let $f \in W^{1,1}_{loc}(\Omega, \mathbf{R}^n)$ be such that $|Df| \in L^n \log^{-1} L(\Omega)$ and $J_f(x) \ge 0$ a.e. Then we have*

$$J_f \in L^1_{loc}(\Omega) \quad and \quad \mathscr{J}_f(\varphi) = \int_\Omega \varphi J_f \quad for \; all \; \varphi \in C^\infty_C(\Omega).$$

It is clear that $J_f \in L^1$ for each $f \in W^{1,n}$ but the previous theorem tells us that, somewhat surprisingly, it is enough to assume that $|Df| \in L^n \log^{-1} L(\Omega)$ if our mapping does not change orientation. This is the key tool in the study of the class of mappings of exponentially integrable distortion. For the proof we need a couple of lemmata.

Lemma 2.13. *Let $\Omega \subset \mathbf{R}^n$ be open and let B be a ball such that $B \subset\subset \Omega$. Suppose that $f \in W^{1,n-\frac{1}{2}}(\Omega, \mathbf{R}^n)$ satisfies $f_1 \in W^{1,\infty}_0(B)$. Then*

$$\int_B J_f = 0.$$

Proof. For smooth functions our claim follows from Proposition 2.10 applied to test functions φ such that $\varphi \equiv 1$ on B since we can extend f_1 as zero outside B.

For a nonsmooth function, pick a sequence of functions $f^k \in C^2$ such that $f_1^k \in W^{1,\infty}_0(B)$,

$$f^k \to f \text{ in } W^{1,n-\frac{1}{2}} \text{ and } Df_1^k \to Df_1 \text{ in } L^{2n-1} .$$

Then we can use a telescopic sum and Hölder's inequality to estimate

$$\left| \int_B (J_{f^k} - J_f) \right| \leq \sum_{i=1}^n \left| \int_B (J(f_1, \ldots, f_{i-1}, f_i^k, \ldots, f_n^k) - J(f_1, \ldots, f_i, f_{i+1}^k, \ldots, f_n^k)) \right|$$

$$\leq C \, \|Df_1^k - Df_1\|_{L^{2n-1}} \||Df|^{n-1}\|_{L^{\frac{n-\frac{1}{2}}{n-1}}} +$$

$$+ C \sum_{i=2}^n \|Df_1\|_{L^{2n-1}} \||Df^k|^{i-2}\|_{L^{\frac{n-\frac{1}{2}}{i-2}}} \|Df_i^k - Df_i\|_{L^{n-\frac{1}{2}}} \||Df|^{n-i}\|_{L^{\frac{n-\frac{1}{2}}{n-i}}}$$

$$\leq C \, \|Df_1^k - Df_1\|_{L^{2n-1}} \|Df\|_{L^{n-\frac{1}{2}}}^{n-1} +$$

$$+ C \sum_{i=2}^n \|Df_1\|_{L^{2n-1}} \|Df_i^k - Df_i\|_{L^{n-\frac{1}{2}}} \|Df\|_{L^{n-\frac{1}{2}}}^{n-2} \xrightarrow{k \to \infty} 0. \qquad \square$$

Given a ball $B(x_0, 3R)$ and a function $v \in L^1(B(x_0, 3R))$ we define

$$M_{3R}v(x) = \sup_{x \in B(y,r) \subset B(x_0, 3R)} \fint_{B(y,r)} |v| \, .$$

Lemma 2.14. *Suppose that* $u \in W^{1,1}(B(x_0, 3R))$ *satisfies* $\int_{B(x_0, R)} u = 0$ *and let* $\lambda > 0$. *There exists* $C = C(n, R) > 0$, *such that for all* $x \in B(x_0, R)$ *we have*

$$M_{3R}u(x) > \lambda \implies M_{3R}|Du|(x) > C\lambda.$$

Proof. We fix x and choose a ball $B_0 \subset B(x_0, 3R)$ containing x such that $M_{3R}u(x) \leq 2|u|_{B_0}$. We proceed similarly to the proof of Lemma A.25 in the appendix, starting from the ball B_0, and we find a sequence of balls $\{B_i\}_{i=0}^k$ with "essentially geometrically increasing" radii r_i such that $B_k = B(x_0, R)$, $\sum_i r_i < CR$ and $|B_i \cap B_{i+1}| \geq C|B_i|$. As $u_{B_k} = 0$ we can proceed similarly to the proof of Lemma A.25 to obtain

$$M_{3R}u(x) \leq 2|u|_{B_0} \leq 2 \fint_{B_0} |u(x) - u_{B_0}| \, dx + 2|u_{B_0}|$$

$$\leq 2 \fint_{B_0} |u(x) - u_{B_0}| \, dx + 2 \sum_{i=0}^{k-1} |u_{B_i} - u_{B_{i+1}}| \leq C(n, R) M_{3R}|Du|(x)$$

and the result follows. $\qquad \square$

Lemma 2.15. *Let* $f \in W^{1,1}_{\text{loc}}(B(x_0, 3R), \mathbf{R}^n)$ *satisfy* $|Df| \in L^n \log^{-1} L(B(x_0, 3R))$, *and let* $\varphi \in C_C^\infty(B(x_0, R))$. *Define*

$$u(x) := \varphi(x) \left(f_1(x) - (f_1)_{B(x_0, R)} \right)$$

and set

$$F_\lambda = \{x \in B(x_0, R) : M_{3R}|Du|(x) < \lambda\}.$$

Then

$$\liminf_{\lambda \to \infty} \lambda \int_{B(x_0, R) \setminus F_\lambda} |Df|^{n-1} = 0.$$

Proof. Write $B = B(x_0, R)$. By the product rule for derivatives we obtain

$$|Du(x)| \le |Df_1(x)||\varphi(x)| + |D\varphi(x)||f_1(x) - (f_1)_B|.$$

Fix $y \in B \setminus F_\lambda$. By integrating the previous inequality over $B(y, r)$ for a fixed $r \in (0, 3R - |y - x_0|)$ we obtain

$$\fint_{B(y,r)} |Du(x)|dx \le \fint_{B(y,r)} |Df_1(x)||\varphi(x)|dx + \fint_{B(y,r)} |D\varphi(x)||f_1(x) - (f_1)_B|dx.$$

$$\le CM_{3R}|Df_1|(y) + CM_{3R}|f_1 - (f_1)_B|(y).$$

Now taking the supremum over r we conclude that

$$\lambda \le M_{3R}|Du|(y) \le CM_{3R}|Df_1|(y) + CM_{3R}|f_1 - (f_1)_B|(y).$$

This implies that at least one of the right-hand side terms is greater than $\frac{\lambda}{2C}$. But this by Lemma 2.14, where we put $u := f_1 - (f_1)_B$, implies that there exists some $C = C(n, R) > 0$ such that $CM_{3R}|Df_1| > \lambda$. Hence

$$B \setminus F_\lambda \subset \{M_{3R}|Df| > C\lambda\}.$$

This however shows, in combination with Lemma A.7, that for any $\delta \in (0, 1)$,

$$\lambda \int_{B \setminus F_\lambda} |Df|^{n-1} \le \lambda \int_{\{M_{3R}|Df| > C\lambda\}} |Df|^{n-1}$$

$$\le C\lambda^{1-\delta} \int_{\{M_{3R}|Df| > C\lambda\}} (M_{3R}|Df|)^{n-1+\delta} \qquad (2.8)$$

$$\le C\lambda^{1-\delta} \int_{\{|Df| > \frac{C\lambda}{2}\}} |Df|^{n-1+\delta}.$$

We now show that the last one of these terms is small. By using the Fubini theorem we get

$$\int_1^\infty \frac{1}{t \log(e + t)} t^{1-\delta} \left(\int_{\{|Df| > t\}} |Df(x)|^{n-1+\delta}dx \right) dt$$

$$= \int_{\{|Df| > 1\}} |Df(x)|^{n-1+\delta} \left(\int_1^{|Df(x)|} \frac{t^{-\delta}}{\log(e + t)} dt \right) dx$$

$$\le C \int_{\{|Df| > 1\}} \frac{|Df(x)|^n}{\log(e + |Df(x)|)} dx < \infty.$$

Since the integrands above are nonnegative and $\int_1^\infty \frac{1}{t \log(e+t)} \, dt = \infty$, we obtain

$$\liminf_{t \to \infty} t^{1-\delta} \left(\int_{\{|Df|>t\}} |Df(x)|^{n-1+\delta} dx \right) = 0 \, . \qquad \qquad \Box$$

Proof (of Theorem 2.12). Let us first show that $J_f \in L^1_{\text{loc}}$. Let $B = B(x_0, R)$ be a ball in Ω such that $3B \subset\subset \Omega$. We choose $\varphi \in C_0^\infty(B)$ such that $\varphi \geq 0$ and $\varphi = 1$ everywhere on $\frac{1}{2}B$. We now define $u(x) = (f_1(x) - (f_1)_B)\varphi(x)$. Clearly $u \in W_0^{1,1}(B)$. Fix $\lambda > 0$. Let us recall the notation of the previous lemma:

$$F_\lambda = \{x \in B : M_{3R}|Du|(x) < \lambda\}.$$

We would like to employ Lemma A.25 to conclude that u is Lipschitz continuous in F_λ. This is strictly speaking not legal since Lemma A.25 deals with the Lebesgue points of u in F_λ. However, the set of Lebesgue points in F_λ has full measure and hence the claim of the preceding lemma also holds for this smaller set that we continue to denote F_λ. This understood, Lemma A.25 gives

$$|u(x) - u(y)| \leq C\lambda|x - y| \qquad \text{for all } x, y \in F_\lambda.$$

We define

$$\tilde{u}_\lambda(x) := \begin{cases} u(x) & x \in F_\lambda \\ 0 & x \in \mathbf{R}^n \setminus B. \end{cases}$$

We claim that this extended function is also $C\lambda$-Lipschitz. By McShane's extension Lemma A.23 we can then extend it to a $C\lambda$-Lipschitz function on the entire set $\mathbf{R}^n$. Now we briefly sketch the proof of this claim. Assume that $x \in F_\lambda$, $y \in \mathbf{R}^n \setminus B$ and set $r = \text{dist}(x, \partial B) \leq |x - y|$. Analogously to the proof of Lemma A.25 in the Appendix we set $B_i = B(x, 4r2^{-i})$ and using Poincaré inequality we obtain

$$|\tilde{u}_\lambda(x)| \leq |u_{B_1}| + \sum_{i=1}^\infty |u_{B_i} - u_{B_{i+1}}| \leq |u_{B_1}| + C \sum_{i=1}^\infty 2^{-i} r M_{3R}|Du|(x)$$

$$\leq |u_{B_1}| + C\lambda|x - y|.$$

For $A := B_1 \cap (\mathbf{R}^n \setminus B)$ we clearly have $|A| \geq C|B_1|$. Since $u \equiv 0$ on A, we obtain analogously to the proof of Lemma A.25 that

$$|u_{B_1}| = \frac{1}{|A|} \left| \int_A (u - u_{B_1}) \right| \leq \frac{1}{|A|} \int_A |u - u_{B_1}|$$

$$\leq \frac{C}{|B_1|} \int_{B_1} |u - u_{B_1}| \leq CrM_{3R}|Du|(x) \leq C\lambda|x - y| \, .$$

It follows that

$$|\tilde{u}_\lambda(y) - \tilde{u}_\lambda(x)| = |\tilde{u}_\lambda(x)| \leq C\lambda |x - y|$$

and hence $\tilde{u}_\lambda$ is Lipschitz with constant $C\lambda$ as claimed. This inequality is obvious for other possible positions of x and y.

Since each function from $L^n \log^{-1} L$ clearly belongs to $L^{n-\frac{1}{2}}$ we obtain from Lemma 2.13 that

$$\int_B J(\tilde{u}_\lambda, f_2, \ldots, f_n) = 0.$$

This, in conjunction with $|D\tilde{u}_\lambda| \leq C\lambda$, gives that

$$\left| \int_{F_\lambda} J(\tilde{u}_\lambda, f_2, \ldots, f_n) \right| = \left| \int_{B \setminus F_\lambda} J(\tilde{u}_\lambda, f_2, \ldots, f_n) \right| \leq C\lambda \int_{B \setminus F_\lambda} |Df|^{n-1}. \quad (2.9)$$

We use the definition of $\tilde{u}_\lambda$ and the product rule for derivatives to get

$$J(\tilde{u}_\lambda, f_2, \ldots, f_n) = \varphi J(f_1, f_2, \ldots, f_n) + (f_1 - (f_1)_B)J(\varphi, f_2, \ldots, f_n) \quad (2.10)$$

almost everywhere on F_λ.

Thus we have, using $J_f \geq 0$, that

$$\int_{F_\lambda \cap \frac{1}{2}B} J_f \leq \int_{F_\lambda} \varphi J_f \leq \left| \int_{F_\lambda} J(\tilde{u}_\lambda, f_2, \ldots, f_n) \right| + \left| \int_{F_\lambda} (f_1 - (f_1)_B)J(\varphi, f_2, \ldots, f_n) \right|.$$

Now using (2.9) and the fact that $|Df| \in L^n \log^{-1} L(\Omega)$ we deduce that

$$\int_{F_\lambda \cap \frac{1}{2}B} J_f \leq C\lambda \int_{B \setminus F_\lambda} |Df|^{n-1} + \int_B |(f_1 - (f_1)_B)J(\varphi, f_2, \ldots, f_n)|$$

for all $\lambda > 0$. Now we may apply $\liminf_{\lambda \to \infty}$ to both sides and by Lemma 2.15 the first term vanishes and the second term can be estimated using Hölder's inequality and Sobolev embedding Theorem A.18 as in Remark 2.11 (a). Together with the nonnegativity of the Jacobian this implies

$$\int_{\frac{1}{2}B} J_f(x)dx = \int_{\frac{1}{2}B} |J_f(x)|dx < \infty.$$

By the arbitrariety of the center x_0 of B, we conclude that $J_f \in L^1_{\text{loc}}(\Omega)$.

Now we shall prove the desired equivalence between J_f and $\mathscr{J}_f$. Notice that at the beginning of the proof we could have taken $\varphi \in C_C^\infty(B)$ without any other restriction, defined $u^\varphi(x) := (f_1(x) - (f_1)_B)\varphi(x)$, and then repeated the extension process as described above to get $\tilde{u}^\varphi_\lambda \in W_0^{1,1}(B)$. Analogously to (2.10) we get

$$\varphi J(f_1, f_2, \ldots, f_n) = J(\tilde{u}^\varphi_\lambda, f_2, \ldots, f_n) - (f_1 - (f_1)_B)J(\varphi, f_2, \ldots, f_n) \quad (2.11)$$

almost everywhere on F_λ and hence

$$\int_B \varphi J_f = \int_{B \setminus F_\lambda} \varphi J_f + \int_{F_\lambda} \varphi J_f = \int_{B \setminus F_\lambda} \varphi J_f + \int_{F_\lambda} J(\tilde{u}^\varphi_\lambda, f_2, \ldots, f_n)$$
$$+ \int_{F_\lambda} (f_1)_B J(\varphi, f_2, \ldots, f_n) - \int_{F_\lambda} f_1 J(\varphi, f_2, \ldots, f_n). \quad (2.12)$$

By Lemma 2.13 and the Lebesgue dominated convergence theorem we have

$$\lim_{\lambda \to \infty} \int_{F_\lambda} (f_1)_B J(\varphi, f_2, \ldots, f_n) = \int_B (f_1)_B J(\varphi, f_2, \ldots, f_n) = 0.$$

Notice that φJ_f is independent of λ and thanks to the continuity of the Lebesgue integral, the first term of the right-hand side of (2.12) tends to zero when λ tends to infinity. Let us now prove that also the second term tends to zero as λ tends to infinity. By Lemma 2.13

$$\int_{F_\lambda} J(\tilde{u}^\varphi_\lambda, f_2, \ldots, f_n) = -\int_{B \setminus F_\lambda} J(\tilde{u}^\varphi_\lambda, f_2, \ldots, f_n)$$

and hence using $|D\tilde{u}^\varphi_\lambda| \leq C\lambda$ we may use Lemma 2.15 to estimate

$$\liminf_{\lambda \to \infty} \left| \int_{F_\lambda} J(\tilde{u}^\varphi_\lambda, f_2, \ldots, f_n) \right| \leq C \liminf_{\lambda \to \infty} \lambda \int_{B \setminus F_\lambda} |Df|^{n-1} = 0. \quad (2.13)$$

We may apply $\liminf$ to (2.12) and by Lebesgue dominated convergence theorem we have

$$\int_B \varphi J_f = -\limsup_{\lambda \to \infty} \int_{F_\lambda} f_1 J(\varphi, f_2, \ldots, f_n) = -\int_B f_1 J(\varphi, f_2, \ldots, f_n)$$

for all $\varphi \in C_C^\infty(B)$. Hence

$$\mathscr{J}_f(\varphi) = -\int_B f_1 J(\varphi, f_2, \ldots, f_n) = \int_B \varphi J(f_1, f_2, \ldots, f_n) \text{ for all } \varphi \in C_C^\infty(B).$$

Let us now consider a general $\varphi \in C_C^\infty(\Omega)$. Take a finite covering of spt φ by balls B_j of the above type and the usual partition of unity ψ_j for these balls. Then

$\varphi = \sum_{j=1}^{k} \psi_j \varphi$ and the claim follows by applying the previous case to $\psi_j \varphi$ and summing up.   □

2.3 Weakly Monotone Mappings

We use Theorem 2.12 to establish a weak notion of monotonicity for a certain class of functions.

Definition 2.16. Let $u \in W^{1,p}(\Omega)$, $p \in [1, \infty)$. Then u is p-weakly monotone if the following holds: For all balls $B \subset\subset \Omega$ and for all $m, M \in \mathbf{R}, m < M$ both of the following implications are satisfied

$$(m - u)^+ \in W_0^{1,p}(B) \Rightarrow \quad u \geq m \text{ a.e. in } B,$$

$$(u - M)^+ \in W_0^{1,p}(B) \Rightarrow \quad u \leq M \text{ a.e. in } B.$$

Let us note that each continuous function f which is p-weakly monotone is actually monotone, i.e. satisfies

$$\operatorname{osc}_B u \leq \operatorname{osc}_{\partial B} u \text{ for all balls } B \subset\subset \Omega .$$

Theorem 2.17. *Let $\Omega \subset \mathbf{R}^n$ be open. Let $f = (f_1, \ldots, f_n)$ have finite distortion and suppose that $|Df| \in L^n \log^{-1} L(\Omega)$. Then $f_1, \ldots, f_n$ are p-weakly monotone for all $p < n$.*

Proof. We will only consider f_1. The proofs for the other component functions are analogous. Fix $M \in \mathbf{R}$ and let B be a ball whose closure lies in Ω. Suppose that $(u - M)^+ \in W_0^{1,p}(B)$. Let us define

$$v := (f_1 - M)^+ \chi_B \text{ and } g := (v, f_2, \ldots, f_n)$$

and choose

$$\varphi \in C_C^\infty(\Omega) \text{ such that } \varphi \geq 0 \text{ and } \varphi(x) = 1 \text{ for } x \in B .$$

Then $v \in W_{\text{loc}}^{1,p}(\Omega)$ for all $p < n$ and $|Dg| \in L^n \log^{-1} L(\Omega)$. By setting

$$E := \{x \in B : f_1(x) > M\},$$

we have

$$J_g = \begin{cases} 0 & \text{a.e. in } B \setminus E, \\ J_f & \text{a.e. in } E. \end{cases}$$

Here the equation holds for such points that are Lebesgue points of the derivative and density points of the respective sets. We use in turn that $J_g \geq 0$, Theorem 2.12, the fact that $v(x) = 0$ for $x \notin B$ and $\nabla\varphi(x) = 0$ for $x \in B$ to conclude that

$$\int_B J_g \leq \int_\Omega \varphi J_g = -\int_\Omega v J(\varphi, f_2, \ldots, f_n) = 0.$$

Since $J_g \geq 0$ it follows that $J_g = 0$ almost everywhere in B, giving $J_f = 0$ almost everywhere in E. Since f has finite distortion it follows that $|Df| = 0$ a.e. on E. Therefore $Df_1 = 0$ a.e. on $\{f_1 > M\}$ yielding $Dv = 0$ a.e. in B. Because $v \in W_0^{1,p}(B)$ we have that $v = 0$ a.e. in B.

The second implication with $(m - f_1)^+$ can be proven analogously.   □

2.4 Oscillation Estimates and Continuity

We proceed by relating our weak monotonicity to actual monotonicity.

Lemma 2.18. *Let $\Omega \subset \mathbf{R}^n$ be open and $p \in [1,\infty)$. Let $u \in W^{1,p}(\Omega)$ be p-weakly monotone in $B(a,R) \subset\subset \Omega$ and $r < R$. Let u_j, $j \in \mathbf{N}$, be the usual convolution approximations of u. For any two Lebesgue points x_0, $y_0 \in B(a,r)$ of u, and for any $\delta > 0$, there is $N \in \mathbf{N}$ such that for all $j > N$ and for all $t \in (r, R)$ we have*

$$|u_j(x_0) - u_j(y_0)| \leq \operatorname{osc}_{S^{n-1}(a,t)} u_j + 2\delta.$$

Proof. It suffices to show that

$$u_j(x_0),\; u_j(y_0) \in (\; \min_{x \in S^{n-1}(a,t)} u_j(x) - \delta,\; \max_{x \in S^{n-1}(a,t)} u_j(x) + \delta)$$

for all $j \in \mathbf{N}$ greater than some N. We prove only $u_j(x_0) < \max_{x \in S^{n-1}(a,t)} u_j(x) + \delta$ for all $j > N$ as the other inequalities are similar. We prove this by contradiction. Suppose that there exist a sequence of natural numbers $\{j_k\}_{k=1}^\infty$ and a sequence of radii $\{t_k\}_{k=1}^\infty \in [r, R]$ such that

$$u_{j_k}(x_0) \geq \max_{x \in S^{n-1}(a,t_k)} u_{j_k}(x) + \delta.$$

Without loss of generality we may assume that t_k converges to some t and that $|t - t_k| < t/2$. Define

$$v_{j_k}(x) := u_{j_k}(x) - u_{j_k}(x_0) + \delta \text{ for } x \in B(a, t_k).$$

Since $v_{j_k}(x) \leq 0$ for all $x \in S^{n-1}(a, t_k)$, we conclude that $(v_{j_k})^+ \in W_0^{1,p}(B(a, t_k))$. Let us define

$$\tilde{v}_{j_k}(x) = v_{j_k}\left(a + (x - a)\tfrac{t_k}{t}\right) \text{ for } x \in B(a, t) .$$

It is easy to see that $(\tilde{v}_{j_k})^+ \in W_0^{1,p}(B(a, t))$. By (the proof of) Theorem A.15 we know that $v_{j_k}(x) \to u(x) - u(x_0) + \delta$ in $W^{1,p}$ because x_0 is a Lebesgue point of u. Therefore

$$\left\| \tilde{v}_{j_k}(x) - \left(u(a + (x - a)\tfrac{t_k}{t}) - u(x_0) + \delta\right) \right\|_{W^{1,p}} \to 0 .$$

It is not difficult to show that

$$\left\| u(x) - u(a + (x - a)\tfrac{t_k}{t}) \right\|_{W^{1,p}} \to 0 .$$

Indeed, this is easy for C^1-functions and for general u it follows by approximation. Therefore we obtain $\tilde{v}_{j_k} \to u - u(x_0) + \delta$ in $W^{1,p}(B(a, R))$. This implies however that $(u - u(x_0) + \delta)^+ \in W_0^{1,p}(B(a, t))$.

Thanks to the weak monotonicity of u we now have,

$$u(x) \leq u(x_0) - \delta$$

for almost all $x \in B(a, t)$. This however cannot be the case as x_0 is a Lebesgue point of u. $\square$

We will need the following well-known version of the Sobolev imbedding theorem.

Lemma 2.19 (Sobolev Imbedding Theorem on Spheres).   *Let $p > n - 1$ and $u \in W^{1,p}(B(0, R))$. Then there is a representative $\hat{u}$ of u such that for almost every $t \in (0, R)$ we have*

$$\operatorname{osc}_{S^{n-1}(0,t)} \hat{u} \leq Ct \left(\fint_{S^{n-1}(0,t)} |Du|^p \right)^{\frac{1}{p}} , \tag{2.14}$$

where $C = C(n, p)$.

Proof. Suppose first that $u \in C^1(B(0, R))$ and $1 = t < R$. We know that $S^{n-1}(0, 1)$ is an $(n-1)$-dimensional space. Since $p > n - 1$ we can apply the usual Sobolev embedding Theorem A.19 to obtain

$$|u(x) - u(y)| \leq C |x - y|^{1 - \frac{n-1}{p}} \left(\int_{S^{n-1}(0,1)} |Du|^p \right)^{\frac{1}{p}} \tag{2.15}$$

for all $x, y \in S^{n-1}(0, 1)$ and hence

$$\operatorname{osc}_{S^{n-1}(0,1)} u \le C_1 \left(\fint_{S^{n-1}(0,1)} |Du|^p \right)^{\frac{1}{p}}.$$

This can be shown analogously to the usual proof in $\mathbf{R}^{n-1}$ or by using a bi-Lipschitz change of variables from a halfsphere to an $(n-1)$-dimensional ball. By scaling of variables we then obtain the analogous inequality on $S^{n-1}(0,t)$ for all $0 < t < R$. The term osc does not scale in this change of variables but $|Du|$ scales by the factor of t and hence we obtain that the multiplicative term in (2.14) equals $C_1 t$.

For a general $u \in W^{1,p}(B(0,R))$ we have an approximating sequence $u_j \in C^1(B(0,R))$ by Theorem A.15. From the proof of Theorem A.15 we see that $Du_j \to Du$ in $L^p(B(0,R))$ and passing to a subsequence we may assume that $u_j \to u$ a.e. By the Fubini theorem and by passing to a subsequence we get that for a.e. $0 < t < R$ we have

$$\int_{S^{n-1}(0,t)} |Du_j|^p \to \int_{S^{n-1}(0,t)} |Du|^p$$

and the integral on the right hand is finite. By an analog of (2.15) for a general t we know that u_j forms an equicontinuous family on $S^{n-1}(0,t)$ and hence there is a subsequence which converges uniformly. Since $u_j \to u$ a.e., we obtain a representative $\hat{u}$ of u. By passing to a limit in (2.14) for u_j we obtain the same inequality for this representative. $\square$

Lemma 2.20. *Let $\Omega \subset \mathbf{R}^n$ be open and $n-1 < p \le n$ for $n > 2$ or $1 \le p \le 2$ for $n = 2$. Suppose that $u \in W^{1,p}(\Omega)$ is p-weakly monotone in $B(a,R) \subset\subset \Omega$ and let $0 < r < R$. Then there is a representative $\hat{u}$ of u such that for almost every $t \in (r,R)$ we have*

$$\operatorname{diam}\left(\hat{u}(B(a,r))\right) \le Ct \left(\fint_{S^{n-1}(a,t)} |D\hat{u}|^p \right)^{\frac{1}{p}}. \tag{2.16}$$

Proof. Let $r \in (0,R)$ and let $x_0, y_0 \in B(a,r)$ be Lebesgue points of u. Using Lemma 2.18 and then (the proof of) Lemma 2.19, for sufficiently large j we have

$$|u_j(x_0) - u_j(y_0)| \le Ct \left(\fint_{S^{n-1}(a,t)} |Du_j|^p \right)^{\frac{1}{p}} + \delta_j \quad \text{for } t \in (r,R), \tag{2.17}$$

where $\delta_j \to 0$ for $j \to \infty$. Because the convolution approximations converge to u in $W^{1,p}(B(a,R))$, it holds that

$$\int_{B(a,r)} |Du_j - Du|^p \to 0.$$

Passing to a subsequence if necessary we may assume that for almost all $t \in [r,R]$

$$\fint_{S^{n-1}(a,t)} |Du_{j_k} - Du|^p \to 0.$$

Fixing t for which the above holds we take the limit over j_k in (2.17) and arrive at

$$|u(x_0) - u(y_0)| \le Ct \left(\fint_{S^{n-1}(a,t)} |Du|^p \right)^{\frac{1}{p}}$$

where C is from Lemma 2.19. It follows that

$$\operatorname{diam}\left(u(\{y \in B(a,r) : y \text{ is a Lebesgue point of } u\})\right) \le Ct \left(\fint_{S^{n-1}(a,t)} |Du|^p \right)^{\frac{1}{p}}.$$

$$(2.18)$$

Now we define a representative of u as

$$\hat{u}(x) := \limsup_{t \to 0} \fint_{B(x,t)} u.$$

Clearly $u = \hat{u}$ at Lebesgue points of u. Using (2.18) it is not difficult to see that (2.16) holds for this representative. $\qquad \square$

Theorem 2.21. *Let $\Omega \subset \mathbf{R}^n$ be open. Suppose that a mapping $f \in W^{1,1}(\Omega, \mathbf{R}^n)$ satisfies $|Df| \in L^n \log^{-1} L_{\mathrm{loc}}(\Omega)$ and suppose that $f_1, \ldots, f_n$ are p-weakly monotone for some $p \in (n-1, n)$. Then f is continuous.*

Proof. Clearly, there exists an increasing convex function $\Phi \in C^\infty(0, \infty)$ and an $M > 0$, such that for all $s > M$ it holds that

$$\Phi(s) = \frac{s^n}{\log(s)}.$$

We can moreover suppose that,

$$\varphi(t) := \Phi(t^{\frac{1}{p}}),$$

is convex. Let $a \in \Omega$ and choose $R > 0$ so that $B(a, R) \subset\subset \Omega$. Let $0 < r < R$. Using Lemma 2.20 and then the Jensen inequality we see that there is a representative $\hat{f}$ of f such that

$$\operatorname{diam} \hat{f}(B(a,r)) \le Ct \left(\varphi^{-1} \circ \varphi \left(\fint_{S^{n-1}(a,t)} |Df|^p \right) \right)^{\frac{1}{p}}$$

$$\le Ct \left(\varphi^{-1} \left(\fint_{S^{n-1}(a,t)} \Phi(|Df|) \right) \right)^{\frac{1}{p}},$$

for a.e. $r < t < R$. Divide by Ct, raise to the power p and apply φ to get

$$\varphi\left(\frac{\operatorname{diam}^p \hat{f}(B(a,r))}{C^p t^p}\right) = \Phi\left(\frac{\operatorname{diam} \hat{f}(B(a,r))}{Ct}\right) \leq \fint_{S^{n-1}(a,t)} \Phi(|Df|).$$

By multiplying by $\omega_{n-1} t^{n-1}$, where ω_{n-1} is the $(n-1)$-dimensional measure of the unit sphere, and then integrating over t from r to R we conclude that

$$\omega_{n-1} \int_r^R \Phi\left(\frac{\operatorname{diam} \hat{f}(B(a,r))}{Ct}\right) t^{n-1} dt \leq \int_{B(a,R)\setminus B(a,r)} \Phi(|Df|)$$

$$\leq \int_{B(a,R)} \Phi(|Df|) < \infty \tag{2.19}$$

because $|Df| \in L^n \log^{-1} L_{\mathrm{loc}}(\Omega)$. The above holds for all $r > 0$. This implies that $\lim_{r \to 0+} \operatorname{diam} f(B(a,r)) = 0$. To show this, let us suppose that the converse is true i.e. that

$$\limsup_{r \to 0+} \operatorname{diam} f(B(a,r)) =: z > 0.$$

Then however, because $\operatorname{diam} f(B(a,r))$ is non-decreasing in r, we have a $\delta > 0$ such that

$$\lim_{r \to 0+} \omega_{n-1} \int_r^R \Phi\left(\frac{\operatorname{diam} f(B(a,r))}{Ct}\right) t^{n-1} dt \geq \lim_{r \to 0+} \omega_{n-1} \int_r^R \Phi\left(\frac{z}{Ct}\right) t^{n-1} dt$$

$$\geq C \int_0^\delta \frac{1}{t \log t^{-1}} dt = \infty$$

which is in contradiction with (2.19). $\square$

Proof (of Theorem 2.3). The claim easily follows from Theorems 2.17 and 2.21.

$\square$

Proof (of Theorem 2.4). Let $x \in \Omega$ and fix $r > 0$ such that $B(x,r) \subset\subset \Omega$. Then $\exp(\lambda K) \in L^1(B(x,r))$ and $J_f \in L^1(B(x,r))$ and by Theorem 2.17, Lemma 2.8 and Theorem 2.21 we have that f is continuous on $B(x,r)$. $\square$

Remark 2.22. The ideas above give us a modulus of continuity for a mapping f of finite distortion when $f \in W^{1,n}$. Indeed, by Lemma 2.20 for $p = n$ we obtain

$$\frac{\operatorname{diam}^n f(B(a,r))}{t} \leq C \fint_{S^{n-1}(a,t)} |Df|^n.$$

By integration over $t \in (r, \sqrt{r})$ we obtain

$$\mathrm{diam}^n f(B(a,r)) \le C \log^{-1} \frac{1}{r} \int_{B(a,\sqrt{r})} |Df|^n \,.$$

2.5 Differentiability Almost Everywhere

Each function in $W^{1,p}$ has an a.e. differentiable representative if $p > n$ and $n \ge 2$. On the other hand there are functions in $W^{1,p}$, $p \le n$, that are not continuous at any point and thus they cannot be differentiable anywhere. Indeed, order all rational point in $\mathbf{R}^n$ (i.e. points that have rational coordinates) by $\{q_k\}_{k=1}^\infty$ and set $u(x) = \sum_{k=1}^\infty 2^{-k} \log^+ \log \frac{1}{|x-q_k|}$. Then $u \in W^{1,n}$ is infinite on a dense set and hence u is nowhere continuous and nowhere differentiable.

We will show that those mappings of finite distortion that belong to $W^{1,n}$ (or satisfy $\exp(\lambda K) \in L^1$) are differentiable a.e.

Theorem 2.23 (Stepanov). *Let $f : \Omega \to \mathbf{R}^m$ be an arbitrary function such that*

$$\limsup_{y \to x} \frac{|f(x) - f(y)|}{|x - y|} < \infty \text{ for a.e. } x \in \Omega \,. \tag{2.20}$$

Then f is differentiable a.e. in Ω.

Proof. We give a proof only for $f : \Omega \to \mathbf{R}$. Let B_j denote all balls with rational centers and with rational radii such that f is bounded on B_j. Denote

$$f_d^j(x) = \sup\{u(x) : u \le f \text{ on } B_j \text{ and } u \text{ is } j - \text{Lipschitz on } B_j\} \text{ and}$$

$$f_u^j(x) = \inf\{u(x) : u \ge f \text{ on } B_j \text{ and } u \text{ is } j - \text{Lipschitz on } B_j\}.$$

The functions f_u^j and f_d^j are j-Lipschitz and therefore differentiable a.e. Pick a point x such that (2.20) holds and all the functions f_u^j, f_d^j (with $x \in B_j$) are differentiable at x. From (2.20) we can find $r > 0$ and $l > 0$ such that

$$|f(x) - f(y)| \le l|x - y| \text{ for } y \in B(x,r) \,. \tag{2.21}$$

We can find $j > l$ such that $x \in B_j \subset B(x,r)$. From (2.21) it is easy to see that $f_u^j(x) = f(x) = f_d^j(x)$. Moreover, both functions f_u^j and f_d^j are differentiable at x and $f_d^j \le f \le f_u^j$. It follows that f is differentiable at x. □

Theorem 2.24. *Let $f \in W_{\mathrm{loc}}^{1,p}(\Omega, \mathbf{R}^n)$, $p > n - 1$ for $n > 2$ or $p \ge 1$ for $n = 2$, be p-weakly monotone and continuous. Then f is differentiable a.e. in Ω.*

Proof. It is not difficult to show that weakly monotone and continuous mappings are monotone and hence

$$\operatorname{osc}_{B(x,r)} f \le \operatorname{osc}_{S^{n-1}(x,r)} f \text{ for all balls } B(x,r) \subset \Omega .$$

Let $B(x_0, 2r) \subset \Omega$. By Lemma 2.19 (Sobolev imbedding theorem on spheres) we obtain, for almost every $r < t < 2r$, that

$$\operatorname{osc}_{B(x_0,r)} f \le \operatorname{osc}_{S^{n-1}(x_0,r)} f \le C t^{1-\frac{n-1}{p}} \left(\int_{S^{n-1}(x_0,t)} |Df|^p \right)^{\frac{1}{p}} .$$

We can choose $r < t < 2r$ such that also

$$\int_{S^{n-1}(x_0,t)} |Df|^p \le \frac{1}{r} \int_{B(x_0,2r) \setminus B(x_0,r)} |Df|^p .$$

It follows that

$$\operatorname{osc}_{B(x_0,r)} f \le C r^{1-\frac{n-1}{p}} \left(\frac{1}{r} \int_{B(x_0,2r) \setminus B(x_0,r)} |Df|^p \right)^{\frac{1}{p}}$$

$$\le C r \left(\frac{1}{r^n} \int_{B(x_0,2r) \setminus B(x_0,r)} |Df|^p \right)^{\frac{1}{p}} .$$

Hence we obtain that at the Lebesgue points of $|Df|^p$ we have

$$\limsup_{r \to 0+} \frac{\operatorname{osc}_{B(x_0,r)} f}{r} < \infty$$

and the claim follows by the Stepanov theorem. □

With the help of Lemma 2.8 and Theorem 2.17 we now easily obtain the following corollary.

Corollary 2.25. *(a) Let $f \in W^{1,p}(\Omega, \mathbf{R}^n)$, $p > n - 1$ for $n > 2$ or $p \ge 1$ for $n = 2$, be a homeomorphism. Then f is differentiable a.e.*
(b) Let $f \in W^{1,n}(\Omega, \mathbf{R}^n)$ be a mapping of finite distortion. Then f is differentiable a.e.
(c) Let $f \in W^{1,1}(\Omega, \mathbf{R}^n)$ be a mapping of finite distortion such that $\exp(\lambda K) \in L^1(\Omega)$ for some $\lambda > 0$. Then f is differentiable a.e.

Remark 2.26. (a) The continuity of $W^{1,n}$-mappings of finite distortion was proven by Gol'dstein and Vodopyanov [39]. The continuity of weakly monotone mappings in $W^{1,n}$ is due to Manfredi [90] and the generalized version for $WL^n \log^{-1} L$ and Theorem 2.4 can be found in the paper of Iwaniec et al. [62].
(b) Theorem 2.12 that deals with the distributional Jacobian is due to Greco [40] and Iwaniec and Sbordone [69]. Already before this it was shown by Müller [99] that $J_f \in L^1 \log L$ if $f \in W^{1,n}$ satisfies $J_f \ge 0$.
(c) Our proof of Stepanov's Theorem 2.23 is due to Malý [88].

(d) The condition $f \in WL^n \log^{-1} L$ is not the most general one that implies
continuity for weakly monotone mappings. It is possible to weaken it further
for example to $f \in WL^n \log^{-1} L \log^{-1} \log L$ and to some finer Orlicz-type
conditions but for example the condition $f \in WL^n \log^{-1-\varepsilon} L$ is not sufficient.
We recommend [62, 73] for further results in this direction.

Chapter 3
Openness and Discreteness

Abstract The aim of this chapter is to study conditions under which a mapping of finite distortion is open (maps open sets to open sets) and discrete (preimage of each point is a discrete set).

3.1 Motivation and Ball's Counterexamples

One of the crucial properties in the models on nonlinear elasticity is that there is no interpenetration of matter. This corresponds to the fact that two parts of the body cannot be mapped to the same place. From the mathematical point of view this means that the map should be one-to-one and thus invertible.

Let us consider the conformal mapping $f(z) = z^2$ in the punctured complex plane which can be identified with punctured $\mathbf{R}^2$. We know that $f \in C^\infty$ is conformal and hence its distortion satisfies $K \equiv 1$. On the other hand each nonzero point has two preimages and this mapping is not invertible. This shows that even for analytically very nice mappings we cannot conclude that the inverse exists without some extra information.

As a first step one usually attempts to conclude that the mapping in question is open and discrete. Note that for example homeomorphisms are automatically open and discrete.

Definition 3.1. Let $\Omega \subset \mathbf{R}^n$ be a domain. We say that the mapping $f : \Omega \to \mathbf{R}^n$ is open if $f(U)$ is open for each open set $U \subset \Omega$. The mapping f is called discrete if the preimage of each point $f^{-1}(y)$ is a discrete set, i.e. it does not have an accumulation point in Ω.

Each open and discrete map which equals to a homeomorphism close to the boundary is necessarily a homeomorphism, see e.g. the proof of Theorem 3.27 below. Moreover, an open and discrete mapping is locally invertible in neighborhoods of most of the points by the following result of Chernavskii [17, 18]. Recall that the branch set of a map is the set of points where it fails to be locally injective.

S. Hencl and P. Koskela, *Lectures on Mappings of Finite Distortion*, Lecture Notes in Mathematics 2096, DOI 10.1007/978-3-319-03173-6_3,
© Springer International Publishing Switzerland 2014

Theorem 3.2. *Let $\Omega \subset \mathbf{R}^n$ be a domain and let $f : \Omega \to \mathbf{R}^n$ be a discrete and open mapping. Then the topological dimension of the branch set B_f satisfies*

$$\dim B_f = \dim f(B_f) \le n - 2 .$$

The following examples show that openness and discreteness may fail even for Lipschitz mappings if the degree of integrability of the distortion is not high enough.

Example 3.3 (Ball). Let $f : (-1, 1)^2 \to \mathbf{R}^2$ be defined by

$$f(x, y) = [x, |x|y] .$$

Then f is not open and discrete since $f^{-1}([0, 0]) = \{0\} \times (-1, 1)$. The derivative of f is

$$Df(x, y) = \begin{pmatrix} 1 & 0 \\ \pm y & |x| \end{pmatrix}$$

for $x \ne 0$ and therefore it is easy to see that f is Lipschitz and $J_f(x, y) = |x| \ge 0$. Hence it is a mapping of finite distortion and its distortion for small enough $|[x, y]|$ equals to

$$K_f(x, y) = \frac{1}{|x|}$$

and it is integrable with any power strictly less than 1.

Analogously, the mapping $f : (-1, 1)^n \to \mathbf{R}^n$ defined as

$$f([x_1, \ldots, x_n]) = [x_1, \ldots, x_{n-1}, \sqrt{x_1^2 + \ldots + x_{n-1}^2} x_n]$$

is a Lipschitz mapping of finite distortion and its distortion for small enough $|x|$ satisfies

$$K_f(x) = \frac{1}{\sqrt{x_1^2 + \ldots + x_{n-1}^2}}$$

and thus $K_f(x) \in L^p$ for every $p < n - 1$. However, f is not open and discrete since $f^{-1}([0, \ldots, 0]) = \{0\}^{n-1} \times (-1, 1)$. Moreover, it is possible to extend this mapping to a Lipschitz mapping $\hat{f} : (-2, 2)^n \to \mathbf{R}^n$ so that the restriction of $\hat{f}$ to a neighborhood of the boundary, $\hat{f}|_{(-2,2)^n \setminus [-1,1]^n}$, is a homeomorphism.

We will prove the following positive results for continuous mappings of finite distortion. Recall that the existence of a continuous representative in this setting follows by Theorems 2.3 and 2.4.

Theorem 3.4. *Let $\Omega \subset \mathbf{R}^n$ be open and let $f \in W^{1,n}_{\mathrm{loc}}(\Omega, \mathbf{R}^n)$ be a continuous mapping of finite distortion such that $K_f \in L^p(\Omega)$ for some $p > n - 1$ or $K_f \in L^1(\Omega)$ for $n = 2$. Then f is either constant or both open and discrete.*

Theorem 3.5. *Let $\Omega \subset \mathbf{R}^n$ be open and let $f : \Omega \to \mathbf{R}^n$ be a continuous mapping of finite distortion. Suppose that there is $\lambda > 0$ such that $\exp(\lambda K_f) \in L^1_{\mathrm{loc}}(\Omega)$. Then f is either constant or both open and discrete.*

3.2 Topological Degree

In the proof of the positive results in this chapter and in the next chapter we will need the concept of topological degree.

Let $f : \Omega \to \mathbf{R}^n$ be a continuous mapping, $U \subset\subset \Omega$ and $y \in \mathbf{R}^n \setminus f(\partial U)$. We would like to define a degree $\deg(y, f, U)$ that somehow corresponds to the number of preimages $f^{-1}(y)$ in U, taking the orientation into account. For $f \in C^1(\Omega, \mathbf{R}^n)$ it is indeed possible to define

$$\deg(y, f, U) = \sum_{x \in f^{-1}(y) \cap U} \operatorname{sgn} J_f(x)$$

if $y \notin f(\partial U)$ and $J_f(x) \neq 0$ for every $x \in f^{-1}(y)$. We need to find some substitute for this that works for every continuous mapping f that belongs to a reasonable Sobolev space. The idea is to use an integral instead of the sum and to define the degree as

$$\int_U \varphi(f(x)) J_f(x) \, dx$$

where φ is an approximation of the Dirac measure at y. We will of course need to show that this definition will not depend on the choice of φ if φ is a smooth function supported in the corresponding component of $\mathbf{R}^n \setminus f(\partial U)$.

Lemma 3.6. *Let $U \subset \mathbf{R}^n$ be a bounded open set. Let $f, g : \overline{U} \to \mathbf{R}^n$ be Lipschitz mappings with $f = g$ on ∂U. Then*

$$\int_U J_f(x) \, dx = \int_U J_g(x) \, dx .$$

Proof. Let us first consider a Lipschitz mapping $h : \overline{U} \to \mathbf{R}^n$ for which a component function h_j vanishes on the boundary. Using McShane's extension, Lemma A.23, we find a mapping $\hat{h} : \mathbf{R}^n \to \mathbf{R}^n$ which equals to h on U, has compact support and satisfies $\hat{h}_j = 0$ on $\mathbf{R}^n \setminus U$. For a large enough ball we may apply Lemma 2.13 to get

$$\int_{\mathbf{R}^n} J_{\hat{h}}(x)\, dx = 0\,.$$

Since $\hat{h}_j = 0$ on $\mathbf{R}^n \setminus U$ we get $J_{\hat{h}} = 0$ on $\mathbf{R}^n \setminus U$ and hence

$$\int_U J_h(x)\, dx = \int_U J_{\hat{h}}(x)\, dx = 0\,.$$

Using telescopic decomposition we obtain

$$
\begin{aligned}
J_f - J_g &= \sum_{i=1}^{n} J(g_1, \ldots, g_{i-1}, f_i, \ldots, f_n) - J(g_1, \ldots, g_i, f_{i+1}, \ldots, f_n) \\
&= J(f_1 - g_1, f_2, \ldots, f_n) + J(g_1, f_2 - g_2, \ldots, f_n) \\
&\quad + \ldots + J(g_1, g_2, \ldots, f_n - g_n)\,.
\end{aligned}
$$

For each of these n terms one coordinate function has zero boundary value and therefore we can apply the previous observation to these mappings to obtain that the integrals over U are zero. By summing up we conclude that

$$\int_U (J_f(x) - J_g(x))\, dx = 0\,.$$

$\square$

Lemma 3.7. *Let $U \subset \mathbf{R}^n$ be a bounded open set. Let $f, g : \overline{U} \to \mathbf{R}^n$ be Lipschitz mappings such that $f = g$ on ∂U. Then, for every smooth function $\varphi \in C^\infty(\mathbf{R}^n)$ we have*

$$\int_U \varphi(f(x)) J_f(x)\, dx = \int_U \varphi(g(x)) J_g(x)\, dx\,.$$

Proof. Let ψ be a C^1-function such that

$$\frac{\partial \psi}{\partial x_1}(y) = \varphi(y) \text{ for every } y \in \mathbf{R}^n\,.$$

From the previous lemma we know that

$$\int_U J(\psi \circ f, f_2, \ldots, f_n)(x)\, dx = \int_U J(\psi \circ g, g_2, \ldots, g_n)(x)\, dx\,. \tag{3.1}$$

We can use the chain rule to conclude

$$\int_U J(\psi \circ f, f_2, \ldots, f_n)(x)\, dx = \sum_{j=1}^{n} \int_U \frac{\partial \psi}{\partial x_j}(f(x)) J(f_j, f_2, \ldots, f_n)(x)\, dx\,.$$

It is easy to see that for all $j \neq 1$ the last Jacobian is zero and hence

$$\int_U J(\psi \circ f, f_2, \ldots, f_n)(x)\, dx = \int_U \varphi(f(x)) J(f_1, f_2, \ldots, f_n)(x)\, dx .$$

Analogously we obtain the same identity for g and with the help of (3.1) we obtain our conclusion. $\square$

Lemma 3.8. *Let $U \subset \mathbf{R}^n$ be a bounded open set. Let $f, g : \overline{U} \to \mathbf{R}^n$ be Lipschitz mappings. Assume that $E \subset \mathbf{R}^n$ is a compact set which does not intersect line segments between $f(x)$ and $g(x)$ for $x \in \partial U$, i.e.*

$$[f(x), g(x)] \cap E = \emptyset \text{ for every } x \in \partial U . \tag{3.2}$$

Then, for every smooth function $\varphi \in C_0^\infty(U)$ with spt $\varphi \subset E$, *we have*

$$\int_U \varphi(f(x)) J_f(x)\, dx = \int_U \varphi(g(x)) J_g(x)\, dx . \tag{3.3}$$

Proof. It is not difficult to check that the set

$$F = \{z \in \overline{U} : [f(z), g(z)] \cap E \neq \emptyset\}$$

is compact and by the assumption (3.2) we have $F \subset U$. Therefore we can find a Lipschitz function $u : \mathbf{R}^n \to [0, 1]$ such that

$$u(x) = \begin{cases} 1 & \text{for } x \in F , \\ 0 & \text{for } x \in \mathbf{R}^n \setminus U . \end{cases}$$

For $x \in \overline{U}$ let us denote

$$h(x) = f(x) + u(x)(g(x) - f(x)) .$$

Then it is easy to check that h is Lipschitz and that $h = f$ on ∂U. By the previous lemma applied to f and h we thus get

$$\int_U \varphi(f(x)) J_f(x)\, dx = \int_U \varphi(h(x)) J_h(x)\, dx$$

for every smooth function $\varphi \in C_0^\infty(E)$. For every x such that $\varphi(h(x)) \neq 0$ we have $h(x) \in E$ since spt $\varphi \subset E$. Since $h(x)$ is a linear combination of $f(x)$ and $g(x)$ we get $x \in F$ by the definition of F. Hence the last integral is equal to the right-hand side of (3.3) since $h = g$ (and thus $J_h = J_g$ a.e.) at each point where $\varphi(h(x)) \neq 0$. $\square$

With the help of the next lemma we will be able to show that the degree is constant on each component of $\mathbf{R}^n \setminus f(\partial U)$.

Lemma 3.9. *Let $U \subset \mathbf{R}^n$ be a bounded open set and let $f : \overline{U} \to \mathbf{R}^n$ be a Lipschitz mapping. Assume that C is a component of $\mathbf{R}^n \setminus f(\partial U)$ and $\varphi \in C_0^\infty(C)$. Then*

$$\int_U \varphi(f(x)) J_f(x) \, dx = a \int_C \varphi(y) \, dy \,,$$

where the constant a does not depend on φ.

Proof. By the additivity of the integral and the density of simple functions in L^1 it is enough to check this identity for $\varphi = \chi_Q$ where $Q \subset C$ is a cube with a rational edge length. Let us fix rational $d > 0$ and let us denote an open cube with center x and edge length d by $Q(x, d)$. Moreover, we fix a cube $Q' \subset\subset C$, we consider the set

$$Q'_d = \{x \in C : Q(x, d) \subset Q'\}$$

and we choose $y, z \in Q'_d$. If y and z are close enough then we can use the previous lemma for f, $g(x) = f(x) - y + z$, $E = Q'$ and approximations of $\chi_{Q(z,d)}$ to conclude that

$$\int_U \chi_{Q(z,d)}(f(x)) J_f(x) \, dx = \int_U \chi_{Q(z,d)}(g(x)) J_g(x) \, dx$$

$$= \int_U \chi_{Q(y,d)}(f(x)) J_f(x) \, dx \,.$$

It follows that the function

$$h(z) = \int_U \chi_{Q(z,d)}(f(x)) J_f(x) \, dx$$

is locally constant and thus constant on Q'_d. Hence we can find a constant $a(d)$ such that

$$\int_U \chi_{Q(z,d)}(f(x)) J_f(x) \, dx = a(d)|Q(0, d)| \text{ for every } z \in Q'_d \,. \qquad (3.4)$$

Next we need to show that this constant does not depend on d. The cube $Q(z, d) \subset Q'$ can be divided into 2^n cubes of edge length d and using (3.4) for $Q(z, d)$ and for these 2^n cubes we easily obtain that

$$a(d)|Q(0, d)| = \int_U \chi_{Q(z,d)}(f(x)) J_f(x) \, dx$$

$$= \sum_{i=1}^{2^n} \int_U \chi_{Q_i}(f(x)) J_f(x) \, dx = 2^n a(\tfrac{d}{2})|Q(0, \tfrac{d}{2})|$$

which implies $a(d) = a(\frac{d}{2})$. Similarly $a(d) = a(\frac{d}{k})$ for every $d \in \mathbf{Q}$, $d > 0$, and every $k \in \mathbf{N}$ and therefore it is easy to see that the value of a does not really depend on d. Moreover, the cube Q' was chosen arbitrarily and thus we can conclude that formula (3.4) holds for a characteristic function of an arbitrary cube $Q \subset C$ with rational edge length.   □

With the help of Lemmas 3.8 and 3.9 we can see that the value of the integral $\int_U \varphi(f) J_f$ does not change if we perturb f close to ∂U or if we choose a different φ. Hence we may consider the usual convolution approximation f_j of our mapping f and by Lemma 3.8 we obtain that for fixed $\varphi \in C_C^\infty(C)$ the value of the integral $\int_U \varphi(f_j(x)) J_{f_j}(x) \, dx$ does not depend on j if j is large enough. This allows us to finally define the topological degree.

Definition 3.10. Let $f : \Omega \to \mathbf{R}^n$ be a continuous mapping and let $U \subset\subset \Omega$ be a domain. Given a component C of $\mathbf{R}^n \setminus f(\partial U)$ let us fix a nonnegative $\varphi \in C_0^\infty(C)$ such that $\int \varphi = 1$. Then the limit

$$\deg(C, f, U) := \lim_{j \to \infty} \int_U \varphi(f_j(x)) J_{f_j}(x) \, dx$$

exists where f_j denote the usual convolution approximation to f and the value of this limit does not depend on the choice of φ. This limit is called the topological degree of f in C with respect to U. For points $y \in C$ we set $\deg(y, f, U) = \deg(C, f, U)$.

Now we show the connection between the degree and the number of preimages.

Lemma 3.11. *Let* $f : \Omega \to \mathbf{R}^n$ *be a continuous mapping,* $U \subset\subset \Omega$ *be a domain and let* C *be a component of* $\mathbf{R}^n \setminus f(\partial U)$. *Then the degree* $\deg(C, f, U)$ *is an integer.*

Proof. It is enough to show that $\int_U \varphi(f) J_f$ is an integer for each smooth f and each $\varphi \in C_0^\infty(C)$ where C is a component of $\mathbf{R}^n \setminus f(\partial U)$ since the general case follows by approximation and the limit of integers must be an integer.

If $J_f = 0$ in $U \cap f^{-1}(C)$ then clearly $\int_U \varphi(f) J_f = 0$. Otherwise $J_f(x_0) \neq 0$ for some $x_0 \in U \cap f^{-1}(C)$ and hence $|C \cap f(U)| > 0$. By the Sard Theorem A.37 we may find $y \in C \cap f(U)$ such that $J_f(x) \neq 0$ for each $x \in f^{-1}(y) \cap U$. Since $f \in C^1(\overline{U})$ it is easy to see that y has only finitely many preimages in U. Let us denote $x_1, \ldots, x_k$ those preimages of y and we set $\varphi(z) = \psi_{\frac{1}{j}}(z - y)$ for j large enough where ψ denotes the smooth kernel from the definition of the convolution (see (A.4) in the Appendix). If j is large enough then $f^{-1}(\mathrm{spt}\,\varphi)$ has exactly k components $V_1, \ldots, V_k$ and we have

$$\int_U \varphi(f(x)) J_f(x) \, dx = \sum_{i=1}^k \int_{V_i} \varphi(f(x)) J_f(x) \, dx \,.$$

Moreover, f is a homeomorphism on a small neighborhood of each x_i (since $J_f(x_i) \neq 0$) and thus f is a homeomorphism on each V_i if j is large enough. By the Area formula (Corollary A.36 (c)) we have

$$\int_{V_i} \varphi(f(x)) J_f(x)\, dx = \begin{cases} \int \varphi(y)\, dy = 1 & \text{if } J_f(x_i) > 0 \\ -\int \varphi(y)\, dy = -1 & \text{if } J_f(x_i) < 0 \end{cases}$$

and our conclusion follows.   □

Let us list some basic properties of the degree that are not difficult to show using the previous lemmata.

Remark 3.12. (a) If $f, g : \Omega \to \mathbf{R}^n$ are continuous, $U \subset\subset \Omega$ and $g = f$ on ∂U, then $\deg(C, f, U) = \deg(C, g, U)$.

(b) If C is a component of $\mathbf{R}^n \setminus f(\partial U)$ such that $C \cap f(U) = \emptyset$, then $\deg(C, f, U) = 0$. This is easy to see since if φ is supported in C then its support does not intersect $f(U)$.

(c) If C is a component of $\mathbf{R}^n \setminus f(\partial U)$ with $\deg(C, f, U) \neq 0$, then $C \subset f(U)$. Indeed, from the part (b) we see that the condition $\deg(C, f, U) \neq 0$ implies $C \cap f(U) \neq \emptyset$. As C is a component of $\mathbf{R}^n \setminus f(\partial U)$ this easily implies $C \subset f(U)$.

(d) If C is the unbounded component of $\mathbf{R}^n \setminus f(\partial U)$, then $\deg(C, f, U) = 0$.

(e) Given a bounded domain U and continuous $f : \partial U \to \mathbf{R}^n$ we can take the Tietze extension $\hat{f} : \mathbf{R}^n \to \mathbf{R}^n$, which is a continuous mapping. For every component C of $\mathbf{R}^n \setminus f(\partial U)$ we can define $\deg(C, f, U) := \deg(C, \hat{f}, U)$ and by (a) this is independent of the choice of $\hat{f}$.

(f) The degree is stable under homotopy. Let $H : \overline{U} \times [0, 1] \to \mathbf{R}^n$ be a continuous mapping such that $H(x, 0) = f(x)$ and $H(1, x) = g(x)$. Moreover, assume that for every $x \in \partial U$ and $t \in [0, 1]$ we have $y \neq H(x, t)$. Then $\deg(y, f, U) = \deg(y, g, U)$.

This follows from Lemma 3.8 if both f and g are Lipschitz. The general case follows by approximation by Lipschitz mappings.

3.3 Topological Degree for Sobolev Mappings

In the proof of openness and discreteness we need to use the fact that a degree of a continuous mappings in a reasonable Sobolev space can be represented as $\int_U \varphi(f) J_f$.

Theorem 3.13. *Let $f \in W^{1,n}(\Omega, \mathbf{R}^n)$ be a continuous mapping. Let $U \subset\subset \Omega$ be a domain and assume that C is a component of $\mathbf{R}^n \setminus f(\partial U)$. Then*

$$\deg(C, f, U) = \int_U \varphi(f(x))J_f(x)\, dx$$

for every nonnegative $\varphi \in C_0^\infty(C)$ which satisfies $\int_C \varphi = 1$.

Proof. From the definition of the degree we know that

$$\deg(C, f, U) := \lim_{j \to \infty} \int_U \varphi(f_j(x))J_{f_j}(x)\, dx , \qquad (3.5)$$

where f_j denote the usual convolution approximations to f. For these convolution approximations, $f_j \to f$ in $W^{1,n}$ and hence $J_{f_j} \to J_f$ in L^1. Moreover, $f_j \to f$ locally uniformly and hence $\varphi(f_j) \to \varphi(f)$ uniformly. It follows that the limit of the right-hand side of (3.5) equals $\int_U \varphi(f(x))J_f(x)\, dx$. $\qquad \square$

Given two vectors $a, b \in \mathbf{R}^n$, we refer to the usual inner product of a and b by $\langle a, b \rangle$.

Lemma 3.14. *Let $f \in W_{\mathrm{loc}}^{1,1}(\Omega, \mathbf{R}^n)$ be a continuous mapping of finite distortion such that $|Df| \in L^n \log^{-1} L(\Omega)$. Then for every $V \in C^1(\overline{f(\Omega)}, \mathbf{R}^n)$ we have*

$$\int_\Omega \langle \mathrm{adj}\, Df(x)(V \circ f)(x), \nabla\varphi(x) \rangle\, dx = -\int_\Omega \mathrm{div}\, V(f(x))J_f(x)\varphi(x)\, dx$$

for every $\varphi \in C_0^\infty(\Omega)$.

Proof. Let us fix $\varphi \in C_0^\infty(\Omega)$. Without loss of generality we may assume that Ω is bounded. By the linearity of our formula we may assume without loss of generality that $V = (v, 0, \ldots, 0)$. Let us define

$$g(x) = (v \circ f(x), f_2(x), \ldots, f_n(x)) .$$

Analogously to the proof of Lemma 3.7 we can use the chain rule to conclude that

$$J_g(x) = \sum_{j=1}^n \frac{\partial v}{\partial y_j}(f(x))J(f_j, f_2, \ldots, f_n)(x) = \frac{\partial v}{\partial y_1}(f(x))J_f(x) .$$

We need an auxiliary function with nonnegative Jacobian and thus we define

$$v^+(y) = v(y) + y_1 \sup_y \left| \frac{\partial v(y)}{\partial y_1} \right| .$$

Now we have $v = v^+ - v^-$, where

$$\frac{\partial v^+(y)}{\partial y_1} \geq 0 \text{ and } \frac{\partial v^-(y)}{\partial y_1} \geq 0 .$$

Again we can use linearity of our formula to assume without loss of generality that $\frac{\partial v}{\partial y_1} \geq 0$ and hence $J_g \geq 0$. It follows that all assumptions of Theorem 2.12 are satisfied and hence $\mathscr{J}_g = J_g$. Now we can use the first line of Laplace's formula for determinants $(A \cdot \text{adj}\, A = I \cdot \det A)$ for the Jacobi matrix of the mapping $(\varphi, f_2, \ldots, f_n)$ and $\mathscr{J}_g = J_g$ to obtain

$$\int_\Omega \langle \text{adj}\, Df(x)(V \circ f)(x), \nabla\varphi(x) \rangle \, dx = \int_\Omega \left\langle \text{adj}\, Df(x) \begin{pmatrix} (v \circ f)(x) \\ \vdots \\ 0 \end{pmatrix}, \nabla\varphi(x) \right\rangle dx$$

$$= \int_\Omega (v \circ f)(x) J(\varphi, f_2, \ldots, f_n)(x) \, dx$$

$$= -\int_\Omega \varphi(x) J_g(x) \, dx$$

$$= -\int_\Omega \varphi(x) \frac{\partial v}{\partial y_1}(f(x)) J_f(x) \, dx$$

$$= -\int_\Omega \text{div}\, V(f(x)) J_f(x)\varphi(x) \, dx. \qquad \square$$

Theorem 3.15. *Let $f \in W^{1,1}_{loc}(\Omega, \mathbf{R}^n)$ be a continuous mapping of finite distortion with $|Df| \in L^n \log^{-1} L(\Omega)$. Let $U \subset\subset \Omega$ be a domain and assume that C is a component of $\mathbf{R}^n \setminus f(\partial U)$. Then*

$$\deg(C, f, U) = \int_U \varphi(f(x)) J_f(x) \, dx$$

for every nonnegative $\varphi \in C_0^\infty(C)$ which satisfies $\int_C \varphi = 1$.

Proof. If $C \cap f(U) = \emptyset$ then we can use Remark 3.12 (b) to see that both sides of the claimed equality are zero. In the remaining part of the proof we assume that $C \cap f(U) \neq \emptyset$. Without loss of generality we may assume that $\text{spt}\,\varphi \subset\subset C$ since the general case follows easily by approximation. We can find $u \in C^2(\mathbf{R}^n)$ such that $\Delta u(x) = \varphi(x)$ (see Theorem A.42 in the Appendix). Let us fix $\psi \in C_0^\infty(U)$ such that $\psi \equiv 1$ on a neighborhood of $f^{-1}(\text{spt}\,\varphi) \cap U$. We have

$$\int_U \varphi(f(x)) J_f(x) \, dx = \int_U \varphi(f(x)) J_f(x)\psi(x) \, dx$$

$$= \int_U (\text{div}\, \nabla u)(f(x)) J_f(x)\psi(x) \, dx$$

$$= -\int_U \langle \text{adj}\, Df(x)\nabla u(f(x)), \nabla\psi(x) \rangle \, dx$$

where we have used the previous lemma in the last identity. Let f_j denote the usual convolution approximations to f. Since $f_j \to f$ in $W^{1,n-1}$ and $f_j \to f$ locally uniformly, we may express the last integral as a limit and then we use the previous lemma for f_j to obtain

$$- \int_U \langle \text{adj}\, Df(x)\nabla u(f(x)), \nabla \psi(x) \rangle\, dx$$

$$= \lim_{j \to \infty} - \int_U \langle \text{adj}\, Df_j(x)\nabla u(f_j(x)), \nabla \psi(x) \rangle\, dx$$

$$= \lim_{j \to \infty} \int_U \Delta u(f_j(x)) J_{f_j}(x)\psi(x)\, dx$$

$$= \lim_{j \to \infty} \int_U \varphi(f_j(x)) J_{f_j}(x)\psi(x)\, dx$$

$$= \lim_{j \to \infty} \int_U \varphi(f_j(x)) J_{f_j}(x)\, dx = \deg(C, f, U)\,. \qquad \square$$

Theorem 3.16. *Let $f \in W^{1,1}_{\text{loc}}(\Omega, \mathbf{R}^n)$ be a continuous mapping of finite distortion with $|Df| \in L^n \log^{-1} L(\Omega)$. Let $U \subset\subset \Omega$ be a domain and assume that C is a component of $\mathbf{R}^n \setminus f(\partial U)$ such that $C \cap f(U) \neq \emptyset$. Then $\deg(C, f, U) > 0$.*

Proof. Let $y \in C \cap f(U)$. Set $V = f^{-1}(C) \cap U$. Since f is continuous and $y \in \mathbf{R}^n \setminus f(\partial U)$, we see that $\overline{f^{-1}(y)} \subset\subset U$. Hence there is a point $x_0 \in V$ that belongs to the boundary of $f^{-1}(y)$. Let $r > 0$ be so small that $B(x_0, r) \subset V$. Suppose that $J_f = 0$ a.e. on $B(x_0, r)$. As f is a mapping of finite distortion this would imply that $|Df| = 0$ a.e. in this set and hence that f maps the entire ball $B(x_0, r)$ to y. This contradicts the choice of x_0 and it follows that $J_f > 0$ on a set of positive measure in V.

By Theorems 2.17 and 2.21 we know that the assumptions of Theorem 2.24 are satisfied and hence f is differentiable almost everywhere. We may now find $y \in C \cap f(U)$ and $x_0 \in \Omega$ such that $f(x_0) = y$, $J_f(x_0) > 0$, f is differentiable at x_0. It follows that

$$\alpha := \inf_{\|z\|=1} |Df(x_0)z| > 0$$

and hence we can find r small enough such that

$$\left| f(x_0 + x) - f(x_0) - Df(x_0)x \right| < r\alpha \quad \text{for every } x \in S^{n-1}(0, r)\,. \qquad (3.6)$$

Let us consider the homotopy

$$H(x, t) = (1 - t)(f(x_0 + x) - f(x_0)) + t Df(x_0)x$$

for $t \in [0, 1]$ and $x \in B(0, r)$. We may apply Remark 3.12 (f) to obtain

$$\deg(0, f(x_0 + x) - f(x_0), B(0, r)) = \deg(0, Df(x_0)x, B(0, r)) . \qquad (3.7)$$

Now we get

$$\deg(C, f, B(x_0, r)) = \deg(y, f, B(x_0, r)) = \deg(0, f(x_0 + x) - f(x_0), B(0, r))$$
$$= \deg(0, Df(x_0)x, B(0, r)) = \operatorname{sgn} J_f(x_0) = +1 .$$

Analogously for almost every $x \in U$ we can find small enough $r > 0$ such that $\deg(C, f, B(x_0, r)) = 1$ if $f(x) = y$ or $\deg(C, f, B(x_0, r)) = 0$ if $f(x) \neq y$. Since $J_f \geq 0$, we easily obtain by Theorem 3.15 that $\deg(C, f, \cdot)$ is nonnegative. Moreover, $\deg(y, f, \cdot)$ is monotone, i.e. for $y \in C \cap f(U)$ and any ball $B \subset U$ with $y \notin f(\partial B)$ we have that $\deg(y, f, B) \leq \deg(y, f, U)$. (To show monotonicity pick φ supported both in the y-component of $\mathbf{R}^n \setminus f(\partial B)$ and the y-component of $\mathbf{R}^n \setminus f(\partial U)$ and apply Theorem 3.15.) Hence $\deg(y, f, U) = \deg(C, f, U) > 0$.
$\square$

Remark 3.17. The previous proof shows that

$$\deg(f(x), f, B(x, r)) = \operatorname{sgn} J_f(x)$$

provided f is differentiable at x with $J_f(x) \neq 0$ and r is small enough. Thus our concept of degree is indeed equivalent to the usual one (see the beginning of Sect. 3.2).

3.4   Proof of Openness and Discreteness

First let us state the main criterion for verification of openness and discreteness.

Theorem 3.18. *Let $f : \Omega \to \mathbf{R}^n$ be a continuous and non-constant mapping of finite distortion such that $|Df| \in L^n \log^{-1} L(\Omega)$. Suppose that $\mathcal{H}^1(f^{-1}(y)) = 0$ for every $y \in \mathbf{R}^n$. Then f is open and discrete.*

Proof. Openness: Let $V \subset \Omega$ be open and let $x \in V$. We know that $\mathcal{H}^1(f^{-1}(f(x))) = 0$ and hence we can find $r > 0$ small enough such that

$$\partial B(x, r) \cap f^{-1}(f(x)) = \emptyset \text{ and } B(x, r) \subset\subset V \subset \Omega .$$

Let us denote by C the $f(x)$-component of $\mathbf{R}^n \setminus f(\partial B(x, r))$. By Theorem 3.16 we have that $\deg(C, f, B(x, r)) > 0$ and by the property of the degree in Remark 3.12 (c) we conclude that $C \subset f(B(x, r))$. This shows that for every $f(x) \in f(V)$ there is an open set C such that $f(x) \in C \subset f(V)$ and hence $f(V)$ is open.

Discreteness: For contradiction, suppose that $f^{-1}(y)$ is not discrete in Ω. It follows that we can find $x \in \Omega$ such that each neighborhood of x contains infinitely many preimages of y. Since $\mathcal{H}^1(f^{-1}(y)) = 0$ we can find $r > 0$ small enough such that

$$\partial B(x,r) \cap f^{-1}(y) = \emptyset \text{ and } B(x,r) \subset\subset \Omega .$$

We denote the y-component of $\mathbf{R}^n \setminus f(\partial B(x,r))$ by C and set $k :=$ $\deg(C, f, B(x,r))$. By Theorem 3.16 we know that $k > 0$. Choose $k + 1$ points in $f^{-1}(y) \cap B(x,r)$ and open ball neighborhoods $U_1, \ldots, U_{k+1}$ of these points such that

$$y \notin f(\partial U_i) \text{ and } U_i \cap U_j = \emptyset \text{ for } i \neq j .$$

By openness we know that the set $\bigcap_{i=1}^{|k|+1} f(U_i)$ is open and hence we can find a smooth nonnegative function $\varphi \in C_0^\infty(C \cap \bigcap_{i=1}^{k+1} f(U_i))$ such that $\int \varphi = 1$. Then Theorems 3.15 and 3.16 give

$$\int_{U_i} \varphi(f(x)) J_f(x) \, dx = \deg(y, f, U_i) \geq 1 .$$

By Theorem 3.15 and nonnegativity of the Jacobian J_f we obtain

$$k = \int_{B(x,r)} \varphi(f(x)) J_f(x) \, dx \geq \sum_{i=1}^{k+1} \int_{U_i} \varphi(f(x)) J_f(x) \, dx$$

$$= \sum_{i=1}^{k+1} \deg(y, f, U_i) \geq k + 1$$

which gives us the desired contradiction.   □

In the proof of Theorem 3.4 we will use convolution approximations and for the approximations we will use the following technical estimate.

Lemma 3.19. *Let f be a mapping in $C^\infty(\Omega, \mathbb{R}^n)$ and let $\Psi \in C^1([0, \infty), [0, \infty))$ with $\Psi(0) = 0$. There is a constant $C = C(n)$ such that for every test-function $\eta \in C_C^\infty(\Omega, [0, \infty))$, we have*

$$\left| \int_\Omega \eta^n(x) \left[n\Psi(|f(x)|^2) + 2|f(x)|^2 \Psi'(|f(x)|^2) \right] J_f(x) \, dx \right|$$

$$\leq C(n) \int_\Omega \eta^{n-1}(x) |\nabla \eta(x)| \, |f(x)| \, \Psi(|f(x)|^2) \, |Df(x)|^{n-1} \, dx. \quad (3.8)$$

Proof. Fix $i \in \{1, \ldots, n\}$. By Proposition 2.10 applied to φ such that $\varphi \equiv 1$ on $\operatorname{spt} \eta$ we have

$$\int_\Omega J\big(f_1, \ldots, f_{i-1}, \eta^n \Psi(|f|^2) f_i, f_{i+1}, \ldots, f_n\big)(x) \, dx = 0. \quad (3.9)$$

Now the chain rule shows that

$$J(f_1, \ldots, f_{i-1}, \Psi(|f|^2), f_{i+1}, \ldots, f_n)(x) = \sum_{j=1}^{n} 2\Psi'(|f|^2) f_j J(f_1, \ldots, f_{i-1}, f_j, f_{i+1}, \ldots, f_n)$$

$$= 2\Psi'(|f|^2) f_i J_f .$$

By the product rule for derivatives applied to (3.9) we thus obtain

$$\int_{\Omega} \eta^n \left[\Psi(|f|^2) + 2\Psi'(|f|^2) f_i^2 \right] J_f$$

$$= - \int_{\Omega} n\eta^{n-1} \Psi(|f|^2) f_i J(f_1, \ldots, f_{i-1}, \eta, f_{i+1}, \ldots, f_n).$$

Summing over i's, we find that

$$\int_{\Omega} \eta^n \left[n\Psi(|f|^2) + 2|f|^2 \Psi'(|f|^2) \right] J_f$$

$$= \sum_{i=1}^{n} - \int_{\Omega} n\eta^{n-1} \Psi(|f|^2) f_i J(f_1, \ldots, f_{i-1}, \eta, f_{i+1}, \ldots, f_n).$$

Thus, the claimed estimate (3.8) follows from the pointwise inequality

$$|J(f_1, \ldots, f_{i-1}, \eta, f_{i+1}, \ldots, f_n)(x)| \le |\nabla \eta(x)| \, |Df(x)|^{n-1}. \qquad \square$$

Proof (of Theorem 3.4). Let $p(n)$ be a fixed number such that $K_f \in L^{p(n)}$, $p(2) = 1$ and $p(n) > n - 1$ for $n \ge 3$. We would like to show that $\mathscr{H}^1(f^{-1}(y)) = 0$ for every $y \in \mathbf{R}^n$ and then apply Theorem 3.18. By the translation $x \to f(x) - y$ it is enough to show this only for $y = 0$, under the assumption that f is not identically equal to zero.

We define $u : \Omega \to \mathbf{R}$ by

$$u(x) = \begin{cases} \log \log(1/|f(x)|) & \text{for } 0 < |f(x)| \le 1/e \\ 0 & \text{otherwise.} \end{cases}$$

For simplicity, we assume in the following that $|f(x)| \le 1/e$. Fix a ball $B \subset 2B \subset\subset \Omega$. We will show that $|Du(x)| \in L^q(B \cap \{|f| > 0\})$ for $q = np(n)/(p(n) + 1)$. To this end, notice first that $u \in L^1_{loc}(\{|f| > 0\})$ and that the chain rule gives

$$|Du(x)| \le \frac{|Df(x)|}{|f(x)| \log(1/|f(x)|)}$$

in $\{|f| > 0\}$. Write $U = 2B \cap \{|f| > 0\}$. The distortion inequality $|Df(x)|^n \le K_f(x)J_f(x)$ together with Hölder's inequality implies that

$$\int_U \frac{|Df|^q}{|f|^q \log^q(1/|f|)} \le \left(\int_U \frac{J_f}{|f|^n \log^n(1/|f|)}\right)^{q/n} \left(\int_U K_f(x)^{p(n)} \, dx\right)^{(n-q)/n},$$

$$(3.10)$$

where $q = np(n)/(p(n) + 1)$. Now it is enough to bound the middle integral. For this, the finite distortion assumption is not a factor and we can apply the standard approximation argument and we can assume that our mapping f is smooth; recall that $f \in W^{1,n}$.

Fix $\varepsilon > 0$ and set

$$\Psi(t) = \frac{1}{2t^{\frac{n}{2}}} \int_0^t \frac{\varphi_\varepsilon(s)}{s \log^n(s^{-1})} \, ds, \quad \text{where } \varphi_\varepsilon(s) = \frac{1}{1 + \varepsilon 2^{\frac{1}{s}}}.$$

For $t > 0$ it is easy to see that

$$n\Psi(t) + 2t\Psi'(t) = \frac{1}{t^{\frac{n}{2}} \log^n(t^{-1})} \varphi_\varepsilon(t)$$

and hence applying Lemma 3.19 with $\eta \in C_C^\infty(2B, [0, \infty))$ we have

$$\int_{2B} \eta^n \frac{J_f}{|f|^n \log^n(|f|^{-2})} \varphi_\varepsilon(|f|^2)$$

$$\le C(n) \int_{2B} |\eta|^{n-1} |\nabla\eta| \left(\frac{|Df|}{|f|}\right)^{n-1} \int_0^{|f|^2} \frac{\varphi_\varepsilon(s)}{s \log^n(s^{-1})} \, ds.$$

Employing the facts that the function $s \to \varphi_\varepsilon(s)$ is increasing and less than or equal to 1, we obtain that

$$\int_{2B} \eta^n \frac{J_f}{|f|^n \log^n(|f|^{-2})} \varphi_\varepsilon(|f|^2) \qquad (3.11)$$

$$\le C(n) \int_{2B} \left[|\eta|^{n-1} |\nabla\eta| \left(\frac{|Df|}{|f|}\right)^{n-1} \varphi_\varepsilon(|f|^2) \int_0^{|f|^2} \frac{ds}{s \log^n(s^{-1})}\right]$$

$$\le C(n) \int_{2B} \left[|\eta|^{n-1} |\nabla\eta| \left(\frac{|Df|}{|f|}\right)^{n-1} [\varphi_\varepsilon(|f|^2)]^{\frac{n-1}{n}} \frac{1}{\log^{n-1}(|f|^{-2})}\right].$$

The terms

$$\frac{\varphi_\varepsilon(|f|^2)}{|f|^n \log^n (|f|^{-2})} \quad \text{and} \quad \frac{[\varphi_\varepsilon(|f|^2)]^{\frac{n-1}{n}}}{|f|^{n-1} \log^{n-1}(|f|^{-2})}$$

are clearly bounded on spt η independently of f and hence the standard approximation argument shows that the last estimate is also valid for $f \in W_{\mathrm{loc}}^{1,n}(2B, \mathbf{R}^n) \cap C(2B, \mathbf{R}^n)$.

The last estimate is thus valid for mappings of finite distortion in our class. Using this estimate, distortion inequality and Hölder's inequality we get

$$\int_{2B} \eta^n \frac{J_f}{|f|^n \log^n (|f|^{-2})} \varphi_\varepsilon(|f|^2)$$

$$\leq C(n) \left(\int_{2B} \eta^n \frac{J_f}{|f|^n \log^n (|f|^{-2})} \varphi_\varepsilon(|f|^2) \right)^{\frac{n-1}{n}} \left(\int_{2B} |\nabla \eta|^n K_f^{n-1} \right)^{\frac{1}{n}}$$

and, so

$$\int_{2B} \eta^n \frac{J_f}{|f|^n \log^n (|f|^{-2})} \varphi_\varepsilon(|f|^2) \leq C(n) \int_{2B} |\nabla \eta|^n K_f^{n-1}.$$

By the Monotone Convergence Theorem, we find that

$$\left| \int_U \eta^n \frac{J_f}{|f|^n \log^n (|f|^{-2})} \right| \leq C(n) \int_{2B} |\nabla \eta|^n K_f^{n-1}. \tag{3.12}$$

It follows from the inequalities (3.10) and (3.12) that

$$\int_{U \cap B} \frac{|Df|^q}{|f|^q \log^q (1/|f|)} \leq C(n,q) \left(\int_{2B} |\nabla \eta|^n K_f^{n-1} \right)^{q/n} \left(\int_{2B} K_f(x)^{p(n)} \, dx \right)^{(n-q)/n} \tag{3.13}$$

where $q = np(n)/(p(n) + 1)$ and η is a compactly supported smooth function so that $\eta = 1$ in B. Note that $q = 1$ if $n = 2$ and $q > n - 1$ if $n \geq 3$.

Let $x \in \partial f^{-1}(0)$ and set $B = (x, \frac{1}{3}d(x, \partial\Omega))$. Then $2B \subset\subset \Omega$. Set $E = f^{-1}(0) \cap B$. We fix $\delta > 0$ and an open nonempty set $F \subset B$ such that $|f| \geq \delta$ on F. Now, for any $k \in \mathbf{N}$ define

$$u_k = \min \left\{ 1, \frac{\log \log(1/|f|) - \log \log(1/\delta)}{k} \right\}.$$

It is easy to check that u_k is continuous, $u_k \geq 1$ on E and $u_k \leq 0$ on F. Now $|D \log \log(1/|f|)| \in L^q(U \cap B)$ implies $\|Du_k\|_{L^q(B)}^q \leq \frac{C}{k^q}$ and we can use Theorem A.39 for $\varepsilon = 1 - n + q$ to obtain

$$\mathcal{H}_\infty^1 \left(f^{-1}(0) \cap \frac{1}{2}B \right) \leq \frac{C}{k^q}.$$

It follows that $\mathcal{H}^1(f^{-1}(0) \cap \frac{1}{2}B) = 0$. This easily shows $\mathcal{H}^1(f^{-1}(0)) = 0$ and hence our claim follows by Theorem 3.18. □

Remark 3.20. If we knew that our mapping f is bounded to one (a.e. point has at most N preimages) then we could obtain the key estimate more easily using the Area formula (see Theorem A.35)

$$\int_{0<|f(x)|<1/e} \frac{J_f(x)}{|f(x)|^n \log^n \frac{1}{|f(x)|}} \, dx \leq N \int_{B(0,1/e)} \frac{dy}{|y^n| \log^n \frac{1}{|y|}} \, dy < \infty .$$

Proof (of Theorem 3.5). This proof is similar to the proof of Theorem 3.4 and therefore we only sketch it and point out the differences. By Lemma 2.8 we know that $|Df| \in L^n \log^{-1} L$ and hence we can use Theorem 3.18. As before we obtain (3.10) and again it is enough to bound the middle integral for smooth approximations f_k.

We obtain (3.11) for smooth functions and now we would like to pass to the limit. This is standard for the right-hand side since $f_k \to f$ in $W^{1,n-1}$. We may apply Theorem 2.12 to obtain that the Jacobian and distributional Jacobian of f agree. This is obviously also true for f_k and hence it is not difficult to show that $J_{f_k} \to J_f$ in L^1. This allows us to pass to the limit also for the left-hand side of (3.11). The rest of the proof is the same as the proof of Theorem 3.4. □

Remark 3.21. Note that we have in fact shown above that each non-constant continuous mapping of finite distortion $f : \Omega \to \mathbf{R}^n$ such that $|Df| \in L^n \log^{-1} L(\Omega)$ and $K_f \in L^p(\Omega)$ for some $p > n - 1$ or $K_f \in L^1(\Omega)$ for $n = 2$ is open and discrete.

3.5 Local Multiplicity of Mappings of Finite Distortion

In this section we prove the following local result.

Theorem 3.22. *Let $f : \Omega \to \mathbf{R}^n$ be a non-constant continuous mapping of finite distortion such that $|Df| \in L^n \log^{-1} L(\Omega)$ and $K_f \in L^p(\Omega)$ for some $p > n - 1$ or $K_f \in L^1(\Omega)$ for $n = 2$. Then, for each compact set $E \subset \Omega$, there is $m \in \mathbf{N}$ such that f is at most m-to-one on E.*

For this proof we need local properties of discrete and open mappings.

Definition 3.23. Let $f : \Omega \to \mathbf{R}^n$ be a continuous mapping. A domain $U \subset\subset \Omega$ is called a normal domain if $f(\partial U) = \partial(f(U))$.

Lemma 3.24. *Let $f : \Omega \to \mathbf{R}^n$ be a continuous and open mapping and let $U \subset\subset \Omega$. Then $\partial(f(U)) \subset f(\partial U)$.*

Proof. Let $y \in \partial(f(U))$, then there are $y_i \in f(U)$ such that $y_i \to y$. We can pick $x_i \in U$ such that $f(x_i) = y_i$. Since U is bounded we may choose a subsequence of x_i that converges to some $x \in \overline{U}$. By continuity we get $f(x) = y$. It remains to show that $x \in \partial U$. Otherwise $x \in U$ and openness of f would imply $y = f(x) \in f(U)$ which is false.   □

Lemma 3.25. *Let* $f : \Omega \to \mathbf{R}^n$ *be a continuous, discrete and open mapping and let* $U' \subset \mathbf{R}^n$ *be a domain. If* U *is a component of* $f^{-1}(U')$ *with* $U \subset\subset \Omega$, *then* U *is a normal domain and* $f(U) = U' \subset\subset f(\Omega)$.

Proof. Let $y \in f(\partial U)$ and pick $x \in \partial U$ such that $f(x) = y$. Since U is a component of $f^{-1}(U')$ we must have $x \notin f^{-1}(U')$ and hence $y = f(x) \notin U'$. In view of $f(U) \subset U'$ this implies $y \notin f(U)$ and so

$$y \in \overline{f(U)} \setminus f(U) = \partial f(U) .$$

This shows $f(\partial U) \subset \partial f(U)$ and together with the previous lemma it implies $f(\partial U) = \partial f(U)$.

Moreover, $f(\partial U) \cap U' = \emptyset$ and so $f(U) = U' \cap f(\overline{U})$. Thus $f(U)$ is closed in U' and by the openness of f also $f(U)$ is open in U'. By the connectivity of U' we have $f(U) = U'$. The claim $f(U) = U' \subset\subset f(\Omega)$ is clear because $\overline{U} \subset \Omega$ is compact.   □

Lemma 3.26. *Let* $f : \Omega \to \mathbf{R}^n$ *be a continuous, discrete and open mapping. For* $x \in \Omega$, *let* $U(x, f, s)$ *be the* x-*component of* $f^{-1}(B(f(x), s))$. *For each* $x \in \Omega$ *there is* $s_x > 0$ *such that* $U(x, f, s)$ *is normal for all* $0 < s < s_x$. *Moreover,*

$$\operatorname{diam} U(x, f, s) \overset{s \to 0+}{\to} 0.$$

Proof. Fix $x \in \Omega$ and choose $r > 0$ so that

$$B(x, r) \subset\subset \Omega \text{ and } f^{-1}(f(x)) \cap \overline{B(x, r)} = \{x\} .$$

Set $s_x = \operatorname{dist}(f(x), f(\partial B(x, r)))$ and let $0 < s < s_x$. Then the x-component of $f^{-1}(B(f(x), s))$ is contained in $B(x, r)$ and the claim follows by the previous lemma. Since f is continuous and $U(x, f, s) \subset B(x, r)$ we easily get

$$\operatorname{diam} U(x, f, s) \overset{s \to 0+}{\to} 0.$$ □

Proof (of Theorem 3.22). From Remark 3.21 we know that f is an open and discrete mapping. Using the previous lemma we can cover E by normal neighborhoods and since E is compact we can suppose that we have a finite covering. We fix one "center" $x \in E$ of this covering. We know that $f(\partial U(x, f, s)) = \partial B(f(x), s)$ and by Theorem 3.16 we obtain $\deg(B(f(x), s), f, U(x, f, s)) > 0$. Let $y \in \mathbf{R}^n$ be arbitrary. We claim that

$$k := \#\{f^{-1}(y) \cap U(x, f, s)\} \le \deg\big(B(f(x), s), f, U(x, f, s)\big) . \tag{3.14}$$

This implies our statement for

$$m := \sum_{\{U(x_i,f,s):\ U(x_i,f,s)\ \text{cover}\ E\}} \deg\big(B(f(x_i),s),\, f,\, U(x_i,f,s)\big) .$$

The claim (3.14) can be shown analogously to the proof of discreteness in Theorem 3.18. The claim is obvious if $f^{-1}(y) \cap U(x,f,s) = \emptyset$. Otherwise we can choose k points in $f^{-1}(y) \cap U(x,f,s)$ and we can choose neighborhoods of these points $U_1,\dots,U_k$ such that

$$y \notin f(\partial U_i) \text{ and } U_i \cap U_j = \emptyset \text{ for } i \neq j .$$

By openness we know that the set $\bigcap_{i=1}^k f(U_i)$ is open and hence we can find a smooth nonnegative function $\varphi \in C_0^\infty(\bigcap_{i=1}^k f(U_i))$ such that $\int \varphi = 1$. By Theorem 3.15, nonnegativity of the Jacobian and Theorem 3.16 we obtain

$$\deg(B(f(x),s),f,U(x,f,s)) = \int_{U(x,f,s)} \varphi(f(x)) J_f(x)\, dx$$

$$\geq \sum_{i=1}^k \int_{U_i} \varphi(f(x)) J_f(x)\, dx = \sum_{i=1}^k \deg(y,f,U_i) \geq k .$$

$\square$

Next we show that mappings in our class that are homeomorphisms close to the boundary are global homeomorphisms.

Theorem 3.27. *Let $\Omega \subset \mathbf{R}^n$ be a bounded domain, $f : \Omega \to \mathbf{R}^n$ be a nonconstant continuous mapping of finite distortion such that $|Df| \in L^n \log^{-1} L(\Omega)$ and $K_f \in L^p(\Omega)$ for some $p > n-1$ or $K_f \in L^1(\Omega)$ for $n = 2$. Assume that there is a compact set $E \subset \Omega$ such that $f : \Omega \setminus E \to f(\Omega) \setminus f(E)$ is a homeomorphism. Then f is a homeomorphism on Ω.*

Proof. We would like to show that f is one-to-one. Pick a domain $U \subset\subset \Omega$ with $E \subset U$ and fix $z \in U \setminus E$. Fix $0 < r < \mathrm{dist}(z, \partial(U \setminus E))$. Then $f(B(z,r))$ is open. Pick a smooth nonnegative function $\varphi \in C_0^\infty(f(B(z,r)))$ with integral 1. Since f is a homeomorphism on $\Omega \setminus E$ and $f(E) \cap f(B(z,r)) = \emptyset$, we may apply Theorems A.35, 3.15 and 3.16 to obtain

$$1 = \int_{f(B(z,r))} \varphi(y)\, dy \geq \int_{B(z,r)} \varphi(f(x)) J_f(x)\, dx$$

$$= \int_U \varphi(f(x)) J_f(x)\, dx = \deg(f(z),f,U) > 0 .$$

Since f is homeomorphic on $\Omega \setminus E$ and $f(E) \cap f(\partial U) = \emptyset$, it follows that $f(U) \setminus f(\partial U)$ has only one component, the $f(z)$-component. Hence $\deg(f(y),f,U) = 1$ for each $y \in U$. Analogously to the proof of Theorem 3.18 we obtain that $f(y)$

has only one preimage. Hence f is continuous and one-to-one on the compact set E and therefore it is homeomorphic also on E. □

Remark 3.28. (a) The counterexample 3.3 is from [8,9].

(b) In our exposition of topological degree we have benefited from [87].

(c) The statement of Theorem 3.4 for $n = 2$ and $K_f \in L^1$ was first shown by Iwaniec and Šverák [70] and for $n > 2$ by Manfredi and Villamor [91]. The generalization to Theorem 3.5 was established by Kauhanen et al. [72]. We used a simpler proof which was later given by Onninen and Zhong [102].

(d) Let us note that the positive results of Theorems 3.4 and 3.5 are not valid under weaker integrability condition on the derivative like $f \in W^{1,p}$ for some $p < n$ or $f \in WL^n \log^\alpha L$ for some $\alpha < -1$; counterexamples are given in [73].

(e) Moreover, if we assume that our mapping f equals to a homeomorphism close to the boundary or that the multiplicity is essentially bounded then the positive result in Theorem 3.4 is valid also in the limiting case $K_f \in L^{n-1}$ (or even $K_f^{n-1}/\log(e + K_f) \in L^1$) for $n \geq 3$. This was shown by Hencl and Malý [54], Hencl and Koskela [48] and it shows that the examples by Ball are optimal.

(f) The positive result may however fail in the limiting case and it was shown by Hencl and Rajala [57] that there is a Lipschitz mapping with $K_f \in L^{n-1}, n \geq 3$, which is not discrete.

(g) Theorems 3.22 and 3.27 are valid not only for mappings of finite distortion in our class. Analogous statements are valid for general continuous, open and discrete mappings but the proofs would require some additional work. We recommend the monograph of Rickman [117, Chap. I.4.] for more information on this subject.

Open problem 4 ([70]). Suppose that $f \in W^{1,n}(\Omega, \mathbf{R}^n), n \geq 3$, is a non-constant mapping of finite distortion such that $K_f \in L^{n-1}(\Omega)$. Is f necessarily open?

Open problem 5. Suppose that $f \in W^{1,n}(\Omega, \mathbf{R}^n), n \geq 3$, is a non-constant mapping of finite distortion such that the inner distortion satisfies $K_I \in L^p(\Omega)$ for some $p > 1$ (see Sect. 7.1. below for the definition of the inner distortion). Is f open and discrete? Let us note that some results in this direction were obtained by Rajala in [111, 112].

Open problem 6. Suppose that $f \in W^{1,n}(\Omega, \mathbf{R}^n), n \geq 3$, is a non-constant mapping of finite distortion such that $K_f \in L^{n-1}(\Omega) \log^\alpha L$ for some $\alpha \in [n - 2, n(n - 2)]$. Is f open and discrete?

If $\alpha < n - 2$ then it is possible to use the counterexample from [57]. If $\alpha > n(n - 2)$ then the answer is in the positive as was shown by Björn [12].

Open problem 7. Suppose that f satisfies the assumptions of Theorem 3.4 and that f is not a constant mapping. Then f is locally bounded to one by Theorem 3.22. Find optimal additional assumptions on Df under which f is actually a local homeomorphism. Some initial results have been obtained by Kovalev and Onninen [85] and Kovalev et al. [86], also see Heinonen and Kilpeläinen [45].

Open problem 8. Let $f : \mathbf{R}^n \to \mathbf{R}^n$, $n \geq 3$, be a non-constant mapping of finite distortion satisfying

$$\limsup_{R \to \infty} R^{-n} \int_{B(0,R)} \varphi(K_f) < \infty$$

for $\varphi(t) = t^p$, $p > n - 1$, or $\varphi(t) = \exp(\lambda t)$ with some $\lambda > 0$. Does f necessarily omit at most finitely many points?

This would be an analog of the Picard theorem that holds besides analytic functions also for mappings of bounded distortion. See [26, 78, 108, 117] for further information.

Open problem 9. Let $f : B(0, 1) \to \mathbf{R}^n$, $n \geq 3$, be a mapping of finite distortion satisfying $\operatorname{diam}(fB(0, 1/2)) = 1$ and

$$\int_{B(0,1)} \exp(\lambda K_f) \leq A.$$

Does there exist $y \in \mathbf{R}^n$ and $r > 0$ depending only on n and A such that f restricted to some domain is a homeomorphism onto $B(y, r)$? See [27, 109, 110] for further information.

Chapter 4
Images and Preimages of Null Sets

Abstract In this chapter we study conditions that guarantee that our mapping maps sets of measure zero to sets of measure zero. We start with the problem in general Sobolev spaces, after which we establish a better result for mappings of finite distortion. Then we introduce a natural class of counterexamples to statements of this type and finally we give a weak condition under which the preimage of a set of measure zero has measure zero for mappings of finite distortion.

4.1 Lusin (N) Condition in Sobolev Spaces

Definition 4.1. Let $\Omega \subset \mathbf{R}^n$ be open. We say that $f : \Omega \to \mathbf{R}^n$ satisfies the Lusin (N) condition if

$$\text{for each } E \subset \Omega \text{ such that } |E| = 0 \text{ we have } |f(E)| = 0 .$$

There are two major motivations for the study of this property. From the physical point of view this property corresponds to the fact that our deformation f of the body in $\mathbf{R}^n$ cannot create new material from "nothing". This would be unnatural in any physically relevant model and hence we would like to know conditions which exclude such pathological behavior.

From the mathematical point of view this property is crucial for the validity of the change of variables formula which is an essential tool in this area. Without this property we only have an inequality for general Sobolev mappings; see Theorem A.35 in Appendix.

We show that the Lusin (N) condition is satisfied for general Sobolev mapping in the supercritical case $p > n$.

Theorem 4.2. *Let $\Omega \subset \mathbf{R}^n$ and $p > n$. Suppose that $f \in W^{1,p}(\Omega, \mathbf{R}^n)$ is continuous. Then f satisfies the Lusin (N) condition.*

S. Hencl and P. Koskela, *Lectures on Mappings of Finite Distortion*, Lecture Notes in Mathematics 2096, DOI 10.1007/978-3-319-03173-6_4,
© Springer International Publishing Switzerland 2014

Proof. From the Sobolev Embedding Theorem (see Theorem A.19 and paragraph after that) we know that for each ball $B = B(x, r) \subset \Omega$ we have

$$\operatorname{osc}_B f \leq C r^{1-\frac{n}{p}} \left(\int_B |Df|^p \right)^{\frac{1}{p}} . \tag{4.1}$$

Let $E \subset \Omega$ be such that $\mathcal{L}_n(E) = 0$, let $\varepsilon > 0$ be fixed and choose an open set $U \supset E$ such that $|U| < \varepsilon$. By the Besicovitch Covering Theorem A.2 we can find countably many balls $B_i \subset U$ such that

$$E \subset \bigcup_i B_i, \quad \mathcal{L}_n\left(\bigcup_i B_i\right) < \varepsilon \quad \text{and} \quad \sum_i \chi_{B_i}(x) \leq C .$$

We obtain

$$\mathcal{L}_n(f(E)) \leq \mathcal{L}_n\left(\bigcup_i f(B_i)\right) \leq \sum_i \mathcal{L}_n\left(f(B_i)\right) \leq C \sum_i \operatorname{osc}_{B_i}^n f .$$

By (4.1) and Hölder's inequality we get

$$\mathcal{L}_n(f(E)) \leq C \sum_i r_i^{(1-\frac{n}{p})n} \left(\int_{B_i} |Df|^p \right)^{\frac{n}{p}}$$

$$\leq C \left(\sum_i r_i^n \right)^{1-\frac{n}{p}} \left(\sum_i \int_{B_i} |Df|^p \right)^{\frac{n}{p}} \tag{4.2}$$

$$\leq C \mathcal{L}_n\left(\bigcup_i B_i\right)^{1-\frac{n}{p}} \left(\int_\Omega |Df|^p \right)^{\frac{n}{p}} \leq C \varepsilon^{1-\frac{n}{p}} \|Df\|_{L^p(\Omega)}^n .$$

By letting $\varepsilon \to 0+$ we obtain our conclusion. □

On the other hand, the Lusin (N) condition may fail for general Sobolev maps in the critical $W^{1,n}$ case.

Theorem 4.3 (Cesari). *Let $n, m \in \mathbf{N}$ with $n \geq 2$. There exists a continuous mapping $f \in W^{1,n}([-1, 1]^n, [-1, 1]^m)$ such that*

$$f([-1, 1] \times \{0\}^{n-1}) = [-1, 1]^m \tag{4.3}$$

and hence f fails the Lusin (N) condition if $m = n$.

Proof. As a basic building block of our construction we need to know that

$$\forall \varepsilon > 0 \; \forall r > 0 \; \exists \tilde{r} \in (0, r) \; \exists h \in W_0^{1,n}(B(0, r)) \cap C(\overline{B(0, r)})$$

$$\text{such that } h(x) = 1 \text{ for all } x \in B(0, \tilde{r}) \text{ and } \|h\|_{W^{1,n}} < \varepsilon . \tag{4.4}$$

Indeed, choose $a > \log\log\frac{1}{r}$ large enough and set

$$h(x) = \min\left\{1, \left(\log(\log\tfrac{1}{|x|}) - a\right)^{+}\right\}.$$

Since $a > \log\log\frac{1}{r}$ it is easy to check that $h \equiv 0$ outside $B(0, r)$. Moreover, h is continuous and $h(x) = 1$ whenever $\log\log\frac{1}{|x|} \geq a + 1$, i.e. on $B(0, e^{e^{-a-1}})$. It remains to estimate

$$\int_{\mathbf{R}^n} |Dh(x)|^n \, dx \leq \int_{B(0, e^{e^{-a}})} \frac{1}{|x|^n \log^n \frac{1}{|x|}} \, dx = C(n) \int_0^{e^{e^{-a}}} \frac{r^{n-1}}{r^n \log^n \frac{1}{r}} \, dr$$

which can be arbitrarily small by choosing sufficiently large a.

By $\mathbb{V}$ we denote the set of the 2^m vertices of the cube $[-1, 1]^m$. The sets $\mathbb{V}^k = \mathbb{V} \times \ldots \times \mathbb{V}$, $k \in \mathbf{N}$, will serve as the sets of indices for our construction. Set $z_0 = \tilde{z}_0 = 0$ and let us proceed by induction. In the first step we divide $[-1, 1]^m$ into 2^m dyadic cubes of edge length 1 and we denote their centers as z_v, $v \in \mathbb{V}$. In the second step we divide each of these cubes into 2^m dyadic cubes of edge length $\frac{1}{2}$ and we obtain $2^m \cdot 2^m$ cubes with centers z_v, $v \in \mathbb{V}^2$, and so on. Formally for $v = [v_1, \ldots, v_k] \in \mathbb{V}^k$ we denote $w = [v_1, \ldots, v_{k-1}]$ and we define

$$z_v = z_w + \frac{1}{2^{k+1}} v_k = z_0 + \frac{1}{2} \sum_{j=1}^{k} \frac{1}{2^j} v_j.$$

Formally we should write $w(v)$ instead of w but for simplicity of notation we will neglect this.

We fix 2^m distinct points c_v, $v \in \mathbb{V}$, on the line segment $c_v \in [-1, 1] \times \{0\}^{n-1}$ and we choose radii r_v such that the balls $\{B_v(c_v, r_v) : v \in \mathbb{V}\}$ are pairwise disjoint. For each $v \in \mathbb{V}$ we use (4.4) to find function a f_v of $W^{1,n}$-norm less than $\frac{1}{4^m}$ such that

$$f_v \in W_0^{1,n}(B_v(c_v, r_v)), \ 0 \leq f_v \leq 1 \text{ and } f_v = 1 \text{ on some small ball } B_v(c_v, \tilde{r}_v).$$

We proceed by induction. For each $w = [v_1, \ldots, v_{k-1}] \in \mathbb{V}^{k-1}$ we have 2^m indices $v = [v_1, \ldots, v_k] \in \mathbb{V}^k$ and we can find pairwise disjoint balls

$$B_v(c_v, r_v) \subset B_w(c_w, \tilde{r}_w) \text{ such that } c_v \in [-1, 1] \times \{0\}^{n-1}.$$

Using (4.4) we now find a function of $W^{1,n}$-norm less than $\frac{1}{4^{km}}$ such that

$$f_v \in W_0^{1,n}(B_v(c_v, r_v)) \ 0 \leq f_v \leq 1 \text{ and } f_v = 1 \text{ on some small ball } B_v(c_v, \tilde{r}_v).$$

Let us set (see Fig. 4.1)

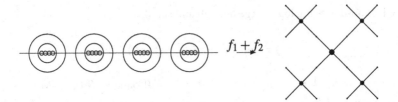

Fig. 4.1 Mapping $f_1 + f_2$ for $n = m = 2$

$$f_k(x) = \sum_{v \in \mathbb{V}^k} f_v(x)(z_v - z_w) .$$

In the k-th step we have 2^{km} functions f_v of $W^{1,n}$-norm less than $\frac{1}{4^{km}}$ and hence

$$\| f_k \|_{W^{1,n}} \leq \frac{2^{km}}{4^{km}} .$$

Moreover, each f_k is continuous and we can estimate

$$\| f_k \|_{L^\infty} \leq C \sup_{v \in \mathbb{V}^k} \| z_v - z_w \|_{\mathbb{R}^m} \leq \frac{C}{2^k}$$

and hence it is easy to see that the limit mapping

$$f(x) = \sum_{k=1}^{\infty} f_k(x) = \sum_{k=1}^{\infty} \sum_{v \in \mathbb{V}^k} f_v(x) \cdot (z_v - z_w)$$

is continuous and satisfies $f \in W^{1,n}([-1, 1]^n, [-1, 1]^m)$. Moreover, for $v \in \mathbb{V}^k$ we have

$$f_k(B_v(c_v, \tilde{r}_v)) = \sum_{i=1}^{k} (z_{(v_1,\ldots,v_i)} - z_{(v_1,\ldots,v_{i-1})}) = z_v .$$

Therefore it is not difficult to find out that the set of all centers z_v is contained in the image,

$$\bigcup_{k \in \mathbb{N}} \bigcup_{v \in \mathbb{V}^k} z_v \subset f([-1, 1] \times \{0\}^{n-1}) .$$

The right-hand side is compact as the image of a compact set under a continuous map and the left-hand side is dense in $[-1, 1]^m$. The conclusion (4.3) follows. $\square$

Remark 4.4. The condition (4.4) means that points have zero capacity with respect to $W^{1,n}$. Note that in the proof above we have only used the condition (4.4) and the rest is independent of the space $W^{1,n}$. In fact, in any function space for which points have zero capacity (for example in $WL^n \log^{n-1} L$) it is possible to construct a similar mapping which violates the Lusin (N) condition.

4.2 Lusin (*N*) Condition for MFD

The aim of this section is to show that mappings of finite distortion satisfy the Lusin (N) condition under weaker assumptions than general Sobolev mappings. Recall that continuous representatives exist by Theorems 2.3 and 2.4.

Theorem 4.5. *Let $\Omega \subset \mathbf{R}^n$ be open and let $f \in W^{1,n}_{\mathrm{loc}}(\Omega, \mathbf{R}^n)$ be a mapping of finite distortion. Then the continuous representative of f satisfies the Lusin (N) condition.*

Theorem 4.6. *Let $\Omega \subset \mathbf{R}^n$ be open and let $f : \Omega \to \mathbf{R}^n$ be a mapping of finite distortion. Suppose that there is $\lambda > 0$ such that $\exp(\lambda K_f) \in L^1_{\mathrm{loc}}(\Omega)$. Then the continuous representative of f satisfies the Lusin (N) condition.*

In the proofs below we employ the following observation.

Lemma 4.7. *Suppose that $f \in W^{1,p}(\Omega, \mathbf{R}^n)$, $p > n - 1$, and let $x \in \Omega$. Then, for almost every $r > 0$ such that the cube $Q(x, r)$ is compactly contained in Ω, the image of its boundary $f(\partial Q(x, r))$ has finite $(n - 1)$-dimensional measure.*

Proof. It is well known (for example from the ACL-condition) that the restriction of a function in $W^{1,p}(\Omega, \mathbf{R}^n)$ to almost all hyperplanes T parallel to coordinate axes belongs to $W^{1,p}(T \cap \Omega, \mathbf{R}^n)$. Hence for almost every $r > 0$, for which $Q(x, r) \subset\subset \Omega$, the restriction of f to $\partial Q(x, r)$ belongs to $W^{1,p}(\partial Q(x, r), \mathbf{R}^n)$. We can use the embedding of $W^{1,p}$, $p > n - 1$, on this $(n - 1)$-dimensional boundary into Hölder continuous functions and analogously to the estimate (4.2) we obtain

$$\mathcal{H}^{n-1}\big(f(\partial Q(x, r))\big) \le C\,\mathcal{H}^{n-1}\big(\partial Q(x, r)\big)^{1 - \frac{n-1}{p}} \left(\int_{\partial Q(x,r)} |Df|^p \right)^{\frac{n-1}{p}} . \qquad \square$$

Theorem 4.8. *Let f be a continuous mapping of finite distortion such that $|Df| \in L^n \log^{-1} L(\Omega)$ and let $E \subset \Omega$ satisfy $|E| = 0$. Then $|f(E)| = 0$.*

Proof. Let $\varepsilon > 0$ and let us fix an open set $\tilde{U} \supset E$ such that $|\tilde{U}| < \varepsilon$. Let $F \subset \tilde{U}$ be a compact set. By Lemma 4.7 we can cover F by cubes $Q_1, \ldots, Q_k \subset \tilde{U}$ such that

$$Q_i \subset \tilde{U}, \quad \sum_{i=1}^{k} |Q_i| < \varepsilon \text{ and } \sum_{i=1}^{k} |f(\partial Q_i)| = 0 . \tag{4.5}$$

Indeed, a set of finite $(n-1)$-dimensional measure necessarily has n-dimensional measure zero.

Let U be a component of $\bigcup_{i=1}^{k} Q_i$. From (4.5) we know $|f(\partial U)| = 0$. Let C_i, $i = 1, \ldots, l$, be components of $\mathbf{R}^n \setminus f(\partial U)$ so that $f(U) \cap C_i \neq \emptyset$ and

$$\sum_{i=1}^{l} |C_i| > \frac{f(U)}{2}.$$

Since $|\partial C_i| = 0$ we can construct functions $\varphi_i \in C_0^\infty(C_i)$ such that $0 \le \varphi_i \le 1$ and $|\{\varphi_i = 1\}| \ge \frac{|C_i|}{2}$. It follows that the function $\tilde{\varphi} = \sum_{i=1}^{l} \varphi_i$ satisfies

$$\operatorname{spt} \tilde{\varphi} \subset \bigcup_{i=1}^{l} C_i \quad \text{and} \quad |\{x \in \mathbf{R}^n : \tilde{\varphi}(x) = 1\}| \ge \frac{|f(U)|}{4}.$$

It easily follows that

$$|f(U)| \le 4|\{x \in \mathbf{R}^n : \tilde{\varphi}(x) = 1\}| \le 4 \int_{\mathbf{R}^n} \tilde{\varphi} = 4 \sum_{i=1}^{l} \int_{\mathbf{R}^n} \varphi_i .$$

Now we may use Theorem 3.16 and then Theorem 3.15 for $\varphi = \varphi_i / \int_{\mathbf{R}^n} \varphi_i$ together with the nonnegativity of the Jacobian to estimate the last sum by

$$|f(U)| \le 4 \sum_{i=1}^{l} \int_{\mathbf{R}^n} \varphi_i \le 4 \sum_{i=1}^{l} \deg(C_i, f, U) \int_{\mathbf{R}^n} \varphi_i$$

$$= 4 \sum_{i=1}^{l} \int_{\Omega} (\varphi_i \circ f) J_f \le 4 \int_U J_f.$$

It follows that

$$|f(F)| \le |f(\bigcup_{i=1}^{k} Q_i)| \le \sum_{\{U: U \text{ is component of } \bigcup_{i=1}^{k} Q_i\}} |f(U)| \le \sum 4 \int_U J_f \le 4 \int_{\tilde{U}} J_f .$$

Taking the supremum over all compact sets $F \subset \tilde{U}$ we obtain $|f(E)| \le 4 \int_{\tilde{U}} J_f$. Since the integral of an integrable function J_f is absolutely continuous with respect to Lebesgue measure, by passing $\varepsilon \to 0$, we obtain $|f(E)| = 0$. $\qquad \square$

Proof (of Theorem 4.5). The claim follows by combining Theorem 4.8 with Theorem 2.3 . $\qquad \square$

Proof (of Theorem 4.6). The claim follows by combining Theorem 4.8 with Lemma 2.8 and Theorem 2.4. $\qquad \square$

4.3 Counterexamples for MFD

First let us recall that Theorem 4.2 holds also in the limiting situation $p = n$ if we moreover assume that our mapping is a homeomorphism.

Theorem 4.9. *Let $\Omega \subset \mathbf{R}^n$ and let $f \in W^{1,n}(\Omega, \mathbf{R}^n)$ be a homeomorphism. Then f satisfies the Lusin (N) condition.*

The proof of this theorem is analogous to the proof of Theorem 1.8 (replace inequality (1.13) by the estimate from Lemma 2.19).

In this section we will show that Theorems 4.5 and 4.6 from the previous section are sharp. First, we show that the Lusin (N) condition may fail even for homeomorphisms if $p < n$ and moreover such a homeomorphism can be chosen to be a mapping of finite distortion.

Theorem 4.10. *Let $p < n$. There exists a homeomorphism of finite distortion $f \in W^{1,p}((-1, 1)^n, (-1, 1)^n)$ for which the Lusin (N) condition fails.*

Proof. We will first give two Cantor-set constructions in $(-1, 1)^n$. Our mapping f will be defined as a limit of a sequence of piecewise continuously differentiable homeomorphisms $f_k : (-1, 1)^n \rightarrow (-1, 1)^n$, where each f_k maps the k-th step of the first Cantor-set construction onto that of the second. Then the limit mapping f maps the first Cantor set onto the second one.

By $\mathbb{V}$ we denote the set of the 2^n vertices of the cube $[-1, 1]^n$. The sets $\mathbb{V}^k = \mathbb{V} \times \ldots \times \mathbb{V}, k \in \mathbf{N}$, will serve as the sets of indices for our construction. Let us denote

$$a_k = \frac{1}{k+1} \quad \text{and} \quad b_k = \frac{1}{2}\left(1 + \frac{1}{k+1}\right). \tag{4.6}$$

Set $z_0 = \tilde{z}_0 = 0$ and let us define

$$r_k = a_k 2^{-k} \quad \text{and} \quad \tilde{r}_k = b_k 2^{-k}. \tag{4.7}$$

It follows that $(-1, 1)^n = Q(z_0, r_0)$ and we proceed by induction. For $v = [v_1, \ldots, v_k] \in \mathbb{V}^k$ we denote $w = [v_1, \ldots, v_{k-1}]$ and we define

$$z_v = z_w + \frac{1}{2}r_{k-1}v_k = z_0 + \frac{1}{2}\sum_{j=1}^{k} r_{j-1}v_j,$$

$$Q'_v = Q(z_v, \tfrac{r_{k-1}}{2}) \quad \text{and} \quad Q_v = Q(z_v, r_k).$$

Formally we should write $w(v)$ instead of w but for simplicity of notation we neglect this.

The number of the cubes in $\{Q_v : v \in \mathbb{V}^k\}$ is 2^{nk}. It is not difficult to show that the resulting Cantor set

Fig. 4.2 *Cubes Q_v and Q'_v*
for $v \in \mathbb{V}^1$ and $v \in \mathbb{V}^2$

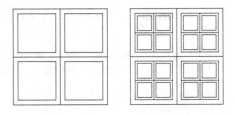

$$\bigcap_{k=1}^{\infty} \bigcup_{v \in \mathbb{V}^k} Q_v =: C_A = C_a \times \ldots \times C_a$$

is a product of n Cantor sets in $\mathbf{R}$ (see Fig. 4.2). Moreover, $\mathscr{L}_n(C_A) = 0$ since

$$\mathscr{L}_n\left(\bigcup_{v \in \mathbb{V}^k} Q_v\right) = 2^{nk}(2a_k 2^{-k})^n \overset{k \to \infty}{\to} 0.$$

Analogously we define

$$\tilde{z}_v = \tilde{z}_w + \frac{1}{2}\tilde{r}_{k-1}v_k = \tilde{z}_0 + \frac{1}{2}\sum_{j=1}^{k}\tilde{r}_{j-1}v_j,$$

$$\tilde{Q}'_v = Q(\tilde{z}_v, \tfrac{\tilde{r}_{k-1}}{2}) \text{ and } \tilde{Q}_v = Q(\tilde{z}_v, \tilde{r}_k).$$

The resulting Cantor set

$$\bigcap_{k=1}^{\infty} \bigcup_{v \in \mathbb{V}^k} \tilde{Q}_v =: C_B = C_b \times \ldots \times C_b$$

satisfies $\mathscr{L}_n(C_B) > 0$ since $\lim_{k \to \infty} b_k > 0$. It remains to find a homeomorphism g which maps C_A onto C_B and satisfies our assumptions. From $\mathscr{L}_n(C_A) = 0$ and $\mathscr{L}_n(C_B) > 0$ we obtain that g does not satisfy the condition (N).

We will give a sequence of homeomorphisms $f_k : (-1, 1)^n \to (-1, 1)^n$. We set $f_0(x) = x$ and we proceed by induction. First we give a mapping f_1 which stretches each cube Q_v, $v \in \mathbb{V}^1$, homogeneously so that $f_1(Q_v)$ equals $\tilde{Q}_v$. On the annulus $Q'_v \setminus Q_v$, f_1 is defined to be an appropriate radial map (radial map in the supremum norm) with respect to z_v and $\tilde{z}_v$ in the image in order to make f_1 a homeomorphism (see Fig. 4.3). The general step is the following: If $k > 1$, f_k is defined as f_{k-1} outside the union of all the cubes Q_w, $w \in \mathbb{V}^{k-1}$. Further, f_k remains equal to f_{k-1} at the centers of the cubes Q_v, $v \in \mathbb{V}^k$. Then f_k stretches each cube Q_v, $v \in \mathbb{V}^k$, homogeneously so that $f_k(Q_v)$ equals $\tilde{Q}_v$. On the annulus $Q'_v \setminus Q_v$, f_k is defined to be an appropriate radial map with respect to z_v in preimage and $\tilde{z}_v$ in image to make f_k a homeomorphism. Notice that the Jacobian determinant $J_{f_k}(x)$ will be strictly positive almost everywhere in $(-1, 1)^n$.

Fig. 4.3 The transformation
of $Q' \setminus Q^\circ$ onto $\tilde{Q}' \setminus \tilde{Q}^\circ$

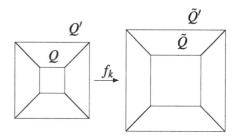

In this proof we use the notation $\|x\|$ for the supremum norm of $x \in \mathbf{R}^n$. The
mappings f_k, $k \in \mathbf{N}$, are formally defined by

$$
f_k(x) = \begin{cases}
f_{k-1}(x) & \text{for } x \notin \bigcup_{v \in \mathbb{V}^k} Q'_v \\
f_{k-1}(z_v) + (\alpha_k \|x - z_v\| + \beta_k)\frac{x - z_v}{\|x - z_v\|} & \text{for } x \in Q'_v \setminus Q_v, \ v \in \mathbb{V}^k \\
f_{k-1}(z_v) + \frac{\tilde{r}_k}{r_k}(x - z_v) & \text{for } x \in Q_v, \ v \in \mathbb{V}^k
\end{cases}
\tag{4.8}
$$

where the constants α_k and β_k are given by

$$
\alpha_k r_k + \beta_k = \tilde{r}_k \quad \text{and} \quad \alpha_k \tfrac{r_{k-1}}{2} + \beta_k = \tfrac{\tilde{r}_{k-1}}{2}.
\tag{4.9}
$$

It is not difficult to check that each f_k is a homeomorphism and maps

$$
\bigcup_{v \in \mathbb{V}^k} Q_v \text{ onto } \bigcup_{v \in \mathbb{V}^k} \tilde{Q}_v.
$$

The limit $f(x) = \lim_{k \to \infty} f_k(x)$ is clearly one-to-one and continuous and therefore
a homeomorphism. Moreover, it is easy to see that f is differentiable almost
everywhere, absolutely continuous on almost all lines parallel to coordinate axes
and maps C_A onto C_B. Furthermore, the pointwise Jacobian of f is strictly positive
a.e. and hence our map is a mapping of finite distortion if its derivative and Jacobian
are locally integrable.

Let $k \in \mathbf{N}$ and $v \in \mathbb{V}^k$. We need to estimate $Df(x)$ in the interior of the annulus
$Q'_v \setminus Q_v$. Since

$$
f(x) = f(z_v) + (\alpha_k \|x - z_v\| + \beta_k)\frac{x - z_v}{\|x - z_v\|}
$$

there, we can almost apply Lemma 2.1 to compute its derivative. The difference
is that instead of the center 0 we use z_v and also that here we have the supremum
norm while in the statement of Lemma 2.1 we have used the usual Euclidean norm.
But these norms are equivalent up to a bi-Lipschitz change and therefore we can
compute the norm of the derivative and the Jacobian using the same rule and the

result is comparable up to a multiplicative constant $C(n)$. From Lemma 2.1, $r_k \approx r_{k-1}$, $\tilde{r}_k \approx \tilde{r}_{k-1}$ (4.9), (4.7) and (4.6) we now obtain

$$Df(x) \approx \max\left\{\frac{\tilde{r}_k}{r_k}, \alpha_k\right\} \approx \max\left\{\frac{b_k}{a_k}, \frac{b_{k-1} - b_k}{a_{k-1} - a_k}\right\} \approx \max\{k, 1\} = k.$$

Moreover, we can estimate

$$\mathscr{L}_n(Q'_v \setminus Q_v) = (r_{k-1})^n - (2r_k)^n \approx 2^{-kn}\left(\frac{1}{(k-1)^n} - \frac{1}{k^n}\right) \approx 2^{-kn}\frac{1}{k^{n+1}}$$

and we have 2^{kn} annuli like that. Therefore, as $p < n$,

$$\int_{Q_0} |Df(x)|^p \, dx \le \sum_{k=1}^{\infty} \sum_{v \in \mathbb{V}^k} \int_{Q'_v \setminus Q_v} |Df(x)|^p \, dx$$

$$\le C \sum_{k=1}^{\infty} 2^{kn} 2^{-kn} \frac{1}{k^{n+1}} k^p < \infty.$$

(4.10)

In conclusion, we have established that f is a homeomorphism, does not satisfy the Lusin (N) condition, that $f \in W^{1,p}((-1, 1)^n, (-1, 1)^n)$ and that $J_f > 0$ almost everywhere. To conclude that f is a mapping of finite distortion, we still need to check that J_f is locally integrable. This follows from the area formula, see Corollary A.36.   □

Let us note that the previous construction is very general and it can be used to construct many other examples. It is enough to plug in different decreasing sequences $\{a_k\}$ and $\{b_k\}$ to obtain new mappings whose properties one can verify in a similar manner.

In this way we can also examine examples for the regularity of the inverse or for the regularity of the composition. It is easy to see that if f is created by using sequences $\{a_k\}$ and $\{b_k\}$, then f^{-1} is created by the same procedure applied to $\{b_k\}$ and $\{a_k\}$. Analogously, if f is created by $\{a_k\}$ and $\{b_k\}$ and g is created by $\{b_k\}$ and $\{c_k\}$, then $g \circ f$ is created by the same construction applied to $\{a_k\}$ and $\{c_k\}$ and we can easily estimate the regularity of the composite mapping.

As another application of this construction we show that the assumptions of Theorem 4.6 are also sharp. Note that for examples constructed in this manner we can also easily estimate the integrability of the distortion function.

Theorem 4.11. *Let $a < 1$. There is a homeomorphism $f \in W^{1,1}((-1, 1)^n, (-1, 1)^n)$ of finite distortion such that $\exp(K_f^a) \in L^1((-1, 1)^n)$, for which the Lusin (N) condition fails.*

Proof. Let us set

$$a_k = \frac{1}{k+1} \quad \text{and} \quad b_k = \frac{1}{2}\left(1 + \frac{1}{\log^{\frac{1}{a}-1}(e+k)}\right).$$

Then we can proceed analogously to the construction from Theorem 4.10. We use (4.7) and the other formulas with these new sequences and we construct two Cantor-type sets in $(-1, 1)^n$. Since $\lim_{k\to\infty} a_k = 0$ and $\lim_{k\to\infty} b_k > 0$ it is easy to see that again the measure of the first Cantor-type set is zero while the measure of the second one is strictly positive.

We construct a sequence of homeomorphisms by using (4.8) (for these new a_k, b_k, r_k, $\tilde{r}_k$). Again we can check that our limit map $f = \lim_{k\to\infty} f_k$ is a homeomorphism, it is differentiable almost everywhere and it satisfies the ACL-condition. It maps the first Cantor-type set onto the second one of strictly positive measure and hence it does not satisfy the Lusin (N) condition. It remains to check the desired integrability of the derivative and distortion; the (local) integrability of the Jacobian of f then follows as in the end of the preceding proof.

Let $k \in \mathbf{N}$ and $v \in \mathbb{V}^k$. It is easy to see that

$$\frac{1}{k} - \frac{1}{k+1} \approx \frac{1}{k^2} \quad \text{and} \quad \frac{1}{\log^{\frac{1}{a}-1}(e+k)} - \frac{1}{\log^{\frac{1}{a}-1}(e+k+1)} \approx \frac{1}{k \log^{\frac{1}{a}}(e+k)} .$$

In the interior of the annulus $Q_v' \setminus Q_v$ we can use Lemma 2.1 to compute

$$|Df(x)| \approx \max\left\{\frac{b_k}{a_k}, \frac{b_{k-1} - b_k}{a_{k-1} - a_k}\right\} \approx k \quad \text{and}$$

$$J_f(x) \approx \left(\frac{b_k}{a_k}\right)^{n-1} \frac{b_{k-1} - b_k}{a_{k-1} - a_k} \approx \frac{k^n}{\log^{\frac{1}{a}}(e+k)}$$

and hence

$$K_f(x) = \frac{|Df(x)|^n}{J_f(x)} \approx \frac{k^n}{\frac{k^n}{\log^{\frac{1}{a}}(e+k)}} = \log^{\frac{1}{a}}(e+k) .$$

Analogously to (4.10) we can compute

$$\int_{Q_0} |Df(x)|\, dx \le C \sum_{k=1}^{\infty} 2^{kn} 2^{-kn} \frac{1}{k^{n+1}} k < \infty \quad \text{and}$$

$$\int_{Q_0} \exp(K_f^a(x))\, dx \le C \sum_{k=1}^{\infty} 2^{kn} 2^{-kn} \frac{1}{k^{n+1}} \exp\left((\log^{\frac{1}{a}}(e+k))^a\right) < \infty .$$

$$(4.11)$$

$\square$

4.4 Lusin (N^{-1}) Condition

In many applications it is also important to know when the preimages of null sets are null sets.

Definition 4.12. Let $\Omega \subset \mathbf{R}^n$ be open. We say that $f : \Omega \to \mathbf{R}^n$ satisfies the Lusin (N^{-1}) condition if

for each $E \subset f(\Omega)$ such that $|E| = 0$ we have $|f^{-1}(E)| = 0$.

The next theorem shows that, for the validity of the (N^{-1}) condition, it is enough to assume that the distortion satisfies $K_f \in L^{\frac{1}{n-1}}$, provided our mapping has essentially bounded multiplicity (number of preimages), i.e. $N(f, \Omega, y) \leq N$ for a.e. $y \in \mathbf{R}^n$. This is, for example, satisfied by homeomorphisms, for which the multiplicity is bounded by one, and locally by discrete and open mappings.

Theorem 4.13. *Let a continuous mapping $f \in W^{1,1}(\Omega, \mathbf{R}^n)$ be a mapping of finite distortion with $K_f^{\frac{1}{n-1}} \in L^1(\Omega)$. If the multiplicity of f is essentially bounded by a constant N and f is not constant, then $J_f(x) > 0$ a.e. in Ω and hence f satisfies the Lusin (N^{-1}) condition.*

We need the following consequence of standard covering arguments.

Lemma 4.14. *There is a constant $\tau = \tau(n)$ with the following property: For each atomless probability Borel measure μ on $\mathbf{R}^n$ there is a point $y \in \mathbf{R}^n$ and a radius $R > 0$ such that*

$$\mu(B(y, 2R)) \geq \tau \quad and \quad \mu(\mathbf{R}^n \setminus B(y, 3R)) \geq \tau.$$

Proof. There is a constant L depending only on n such that any ball with radius $3r$ can be covered by L balls with radius r. Set

$$\tau = \frac{1}{L + 1}.$$

Let

$$\rho = \inf\{r > 0 : [\exists y \in \mathbf{R}^n : \mu(B(y, r)) \geq \tau]\}.$$

Since μ does not have atoms and $\mu(\mathbf{R}^n) = 1$, it easily follows that $\rho > 0$. We find a point $y \in \mathbf{R}^n$ and a radius $R > 0$ such that

$$\mu(B(y, 2R)) \geq \tau \quad and \quad R < \rho < 2R.$$

Consider a collection $B_1, \ldots, B_L$ of balls with radii R that cover $B(y, 3R)$. Since $R < \rho$, we have

$$\mu(B_i) \leq \tau, \qquad i = 1, \ldots, L$$

and thus

$$\mu(B(y, 3R)) \le \sum_{i=1}^{L} \mu(B_i) \le L\tau,$$

which gives, since $\mu(\mathbf{R}^n) = 1$, that

$$\mu(\mathbf{R}^n \setminus B(y, 3R)) \ge 1 - L\tau = \tau. \qquad \square$$

Proof (of Theorem 4.13). The proof is divided into four steps.

STEP 1.   We first prove an auxiliary estimate. Let $E \subset \Omega$ be a measurable set. Consider a smooth function u with a compact support in $\mathbf{R}^n$. Using the distortion inequality, Hölder's inequality and the Area formula (A.18) we obtain

$$\int_E |D(u \circ f)| \, dx \le \int_E |(\nabla u) \circ f| \, |Df| \, dx \le \int_E |(\nabla u) \circ f| \, J_f^{1/n} K^{1/n} \, dx$$

$$\le \left(\int_E |(\nabla u) \circ f|^n \, J_f \, dx \right)^{1/n} \left(\int_E K^{n'-1} \, dx \right)^{1/n'}$$

$$\le \left(N \int_{\mathbf{R}^n} |\nabla u|^n \, dx \right)^{1/n} \left(\int_E K^{n'-1} \, dx \right)^{1/n'}. \qquad (4.12)$$

STEP 2.   We claim that

$$y_0 \in \mathbf{R}^n \implies |f^{-1}(\{y_0\})| = 0. \qquad (4.13)$$

Towards this, consider an arbitrary ball $B \subset\subset \Omega$ of radius $r > 0$ and $y_0 \in \mathbf{R}^n$. Suppose that f differs from y_0 on a set of positive measure in B. Then there is an $R > 0$ such that

$$\kappa := \left| B \setminus f^{-1}(B(y_0, R)) \right| > 0. \qquad (4.14)$$

We know that points have zero n-capacity, i.e. analogously to (4.4) for any given $\varepsilon > 0$ there is a smooth function u on $\mathbf{R}^n$ such that

$$\text{spt } u \subset B(y_0, R), \quad u(y_0) = 1 \quad \text{and} \quad \int_{\mathbf{R}^n} |\nabla u|^n \, dy < \varepsilon^n. \qquad (4.15)$$

Then

$$\min\{|B \cap f^{-1}(y_0)|, \, \kappa\} \le C \, r \int_B |D(u \circ f)| \, dx. \qquad (4.16)$$

For this we used the well-known trick

$$\frac{1}{2} \min\{|B \cap \{v \geq 1\}|, \ |B \cap \{v \leq 0\}|\} \leq \inf_{c \in \mathbf{R}} \int_B |v - c| \, dx \leq C \, r \int_B |Dv| \, dx,$$

$$(4.17)$$

based on the Poincaré inequality, where the hypothesis is that $v \in W^{1,1}(B)$. By (4.12), (4.15) and (4.16) we have

$$\min\{|B \cap f^{-1}(y_0)|, \ \kappa\} \leq C \, r \left(N \int_{\mathbf{R}^n} |\nabla u|^n \, dx \right)^{1/n} \left(\int_B K^{n'-1} \, dx \right)^{1/n'}$$

$$\leq C\varepsilon \left(\int_B K^{n'-1} \, dx \right)^{1/n'}.$$

Letting $\varepsilon \to 0$ and using (4.14) we obtain that $|B \cap f^{-1}(y_0)| = 0$ whenever f differs from y_0 on a set of positive measure in B. Hence (4.13) follows by taking the connectedness of Ω and the assumption that f is not constant into account.

STEP 3. Now, let us prove that $J_f > 0$ a.e. We write $Z = \{x \in \Omega : J_f(x) = 0\}$. Fix a ball $B(x_0, r) \subset \Omega$ and consider the Borel probability measure defined by

$$\mu(A) = \frac{|B(x_0, r) \cap f^{-1}(A)|}{|B(x_0, r)|}, \quad A \subset \mathbf{R}^n.$$

By (4.13), μ does not have atoms. Using Lemma 4.14 we find a point $y \in \mathbf{R}^n$ and a radius $R > 0$ such that

$$\mu(B(y, 2R)) \geq \tau \quad \text{and} \quad \mu(\mathbf{R}^n \setminus B(y, 3R)) \geq \tau. \tag{4.18}$$

where $\tau = \tau(n) > 0$. We can easily find a smooth function u on $\mathbf{R}^n$ such that

$$\text{spt}\, u \subset B(y, 3R), \qquad u = 1 \text{ on } B(y, 2R), \qquad \int_{\mathbf{R}^n} |Du|^n \, dy \leq C(n). \tag{4.19}$$

Set

$$v := u \circ f.$$

Then $v \in W^{1,1}(B(x_0, r))$ and by (4.18) and (4.19),

$$\frac{|B(x_0, r) \cap \{v = 1\}|}{|B(x_0, r)|} \geq \frac{|B(x_0, r) \cap f^{-1}(B(y, 2R))|}{|B(x_0, r)|} = \mu(B(y, 2R)) \geq \tau,$$

$$\frac{|B(x_0, r) \cap \{v = 0\}|}{|B(x_0, r)|} \geq \frac{|B(x_0, r) \setminus f^{-1}(B(y, 3R))|}{|B(x_0, r)|} = \mu(\mathbf{R}^n \setminus B(y, 3R)) \geq \tau.$$

$$(4.20)$$

By the Poincaré inequality (4.17) we have

$$1 \leq Cr^{-n} \inf_{c \in \mathbf{R}} \int_{B(x_0,r)} |v - c| \, dx \leq Cr^{1-n} \int_{B(x_0,r)} |Dv| \, dx$$

$$= Cr^{1-n} \int_{B(x_0,r)} |D(u \circ f)| \, dx. \tag{4.21}$$

Since f is a mapping of finite distortion, we have $Df = 0$ and thus $\nabla(u \circ f) = 0$ a.e. on Z. Hence by (4.21), (4.12) and (4.19)

$$1 \leq Cr^{1-n} \int_{B(x_0,r) \setminus Z} |D(u \circ f)| \, dx \leq C \left(r^{-n} \int_{B(x_0,r) \setminus Z} K^{n'-1} \right)^{1/n'}.$$

If x_0 is a Lebesgue point for $g := K^{n'-1} \chi_{\Omega \setminus Z}$, where $\chi_{\Omega \setminus Z}$ is the characteristic function of $\Omega \setminus Z$, it follows that $g(x_0) > 0$. This means that the set Z does not contain any Lebesgue points of g, and so it must be of measure zero. We have proved that $J_f > 0$ a.e.

STEP 4.   It remains to show the (N^{-1}) condition. Given $E \subset \Omega$ with $|f(E)| = 0$, we find a Borel measurable set $A \subset \mathbf{R}^n$ of measure zero which contains $f(E)$. Then E is contained in the measurable set $E' = f^{-1}(A)$. Let h be the characteristic function of A. By (A.18) we have

$$\int_{E'} J_f(x) \, dx = \int_{\Omega} h(f(x)) \, J_f(x) \, dx \leq N \int_{\mathbf{R}^n} h(y) \, dy = 0.$$

Since $J_f > 0$ a.e. it follows that $|E| = 0$. $\square$

Moreover, it is possible to show that the assumptions on the integrability of the distortion in Theorem 4.13 are sharp.

Theorem 4.15. *Let $a < \frac{1}{n-1}$. There exists a Lipschitz homeomorphism f of finite distortion such that $f \in W^{1,1}((-1,1)^n, (-1,1)^n)$ and $K_f^a \in L^1((-1,1)^n)$, for which the Lusin condition (N^{-1}) fails.*

Proof. Let us set

$$a_k = \frac{1}{2} \left(1 + \frac{1}{k+1} \right) \text{ and } b_k = \frac{1}{k+1} .$$

Again we can use the general construction from Theorem 4.10. In fact we will construct a mapping which is inverse to the mapping constructed in the proof of Theorem 4.10. We use (4.7) and the other formulas with these new sequences and we construct two Cantor-type sets in $(-1,1)^n$. Since $\lim_{k \to \infty} a_k > 0$ and $\lim_{k \to \infty} b_k = 0$ it is easy to see that the measure of the first Cantor-type set is strictly positive while the measure of the second one is zero.

We construct a sequence of homeomorphisms by using (4.8) (for these new a_k, b_k, r_k, $\tilde{r}_k$). Again we can check that our limit map $f = \lim_{k\to\infty} f_k$ is a homeomorphism. Moreover, we will see that all f_k are Lipschitz with constant which does not depend on k. It follows from the construction that f is also Lipschitz. Moreover, it maps the first Cantor-type set of strictly positive measure onto the second one of measure zero and hence it does not satisfy the Lusin (N^{-1}) condition. Let us check the uniform boundedness of the derivative and the integrability of the distortion.

In the interior of the annulus $Q'_v \setminus Q_v$ we can use Lemma 2.1 to compute

$$|Df(x)| \approx \max\left\{\frac{b_k}{a_k}, \frac{b_{k-1} - b_k}{a_{k-1} - a_k}\right\} \approx \max\left\{\frac{1}{k}, 1\right\} \approx 1$$

and

$$J_f(x) \approx \left(\frac{b_k}{a_k}\right)^{n-1} \frac{b_{k-1} - b_k}{a_{k-1} - a_k} \approx \frac{1}{k^{n-1}}$$

and hence

$$K_f(x) = \frac{|Df(x)|^n}{J_f(x)} \approx k^{n-1}.$$

It follows that f is Lipschitz and analogously to (4.11) we can compute with the help of $a < \frac{1}{n-1}$ that

$$\int_{Q_0} K_f^a(x)\, dx \leq C \sum_{k=1}^{\infty} 2^{kn} 2^{-kn} a_k^{n-1} (a_k - a_{k+1}) k^{(n-1)a} \leq C \sum_{k=1}^{\infty} \frac{1}{k^2} k^{(n-1)a} < \infty.$$

The first Cantor-type set is mapped onto the second one of zero volume and hence we can easily deduce from the Area formula (A.18) that $J_f = 0$ a.e. on this set of positive measure. It follows from (4.8) that

$$|Df_k| \leq C \frac{\tilde{r}_k}{r_k} = C \frac{b_k}{a_k}$$

on the k-th iteration of the Cantor set and hence also on the limiting Cantor set. Since $\lim_{k\to\infty} \frac{b_k}{a_k} = 0$ we obtain that the derivative of the limit f of mappings f_k satisfies $|Df(x)| = 0$ a.e. in the first Cantor-type set. It is not difficult to check now that f is a mapping of finite distortion. $\square$

Remark 4.16. (a) The Lusin (N) condition for continuous Sobolev mappings in $W^{1,p}$, $p > n$, was established by Marcus and Mizel [92].

(b) The idea of the construction in the proof of Theorem 4.3 goes back to Cesari [16] and it was further refined by Malý and Martio [89]. A detailed proof of Theorem 4.9 can also be found in [89], see also [114].

(c) The positive results for mapping of finite distortion from Sect. 4.2 are due to Kauhanen et al. [71].
(d) The construction of a homeomorphism without the Lusin condition (N) was first given by Ponomarev [105]. In Sect. 4.3 we have applied it to mappings of finite distortion as was done in [71].
(e) The proof of the (N^{-1}) condition for mappings of finite distortion was given by Koskela and Malý [76].
(f) It is even possible to construct a homeomorphism $f \in W^{1,p}([0,1]^n, [0,1]^n)$, $p < n$, such that $J_f = 0$ a.e. [47], but the construction is more demanding. It follows that $[0,1]^n = A \cup B$ where

$$\mathcal{L}_n(A) = \mathcal{L}_n([0,1]^n), \text{ but } \mathcal{L}_n(f(A)) = 0 \text{ and}$$
$$\mathcal{L}_n(B) = 0, \text{ but } \mathcal{L}_n(f(B)) = \mathcal{L}_n(f([0,1]^n)) .$$

For further refinements see [15, 23].

Open problem 10. We would like to understand the images of smaller null sets. A model problem is: Does there exists a homeomorphism of locally exponentially integrable distortion in the plane so that it maps a line segment L to a line segment L and a set $E \subset L$ with $\mathcal{H}^1(E) > 0$ to a copy of the usual ternary Cantor set?

It is known that already quasiconformal (or quasisymmetric) mappings can map a set $E \subset L$ with $\mathcal{H}^1(E) > 0$ to a set of dimension close to zero, but the image of such a set E under a quasiconformal mapping cannot be a copy of the usual ternary Cantor set, because this set is uniformly porous.

For the properties of the images of sets of Hausdorff dimension strictly less than the underlying dimension n, see [14, 83, 113, 121].

Open problem 11. In Theorem 4.9 we have seen that $W^{1,n}$-homeomorphisms satisfy the Lusin (N) condition. The proof in [89] actually works for continuous monotone (i.e. $\mathrm{osc}_B \, f \leq \mathrm{osc}_{\partial B} \, f$) mappings in $W^{1,n}$. On the other hand the proof of Theorem 4.8 gives the Lusin condition (N) for homeomorphisms in $WL^n \log^{-1} L$, see [71]. Is it true that the Lusin condition (N) holds for continuous monotone mappings in $WL^n \log^{-1} L$? Notice that the homeomorphic counterexamples that we have discuss above, map a rather nice (even a porous) Cantor set onto a set of positive volume. For such sets, the image under a continuous monotone mappings in $WL^n \log^{-1} L$ is necessarily of volume zero [84].

Open problem 12. Recall from Theorem 4.3 that the Lusin condition (N) may fail for a general continuous mapping f that belongs to $W^{1,n}$. On the other hand, if f is additionally Hölder continuous, then condition (N) holds by results in [89]. What is the optimal modulus of continuity for mappings in $W^{1,n}$ that guarantees the Lusin condition (N)? For partial progress on this problem see [77].

Chapter 5
Homeomorphisms of Finite Distortion

Abstract In this chapter we establish the optimal regularity of the inverse mapping in higher dimensions and optimal Sobolev regularity for composites. Moreover, we establish optimal moduli of continuity for mappings in our classes and we discuss orientation preservation and approximation of Sobolev homeomorphisms.

5.1 Regularity of the Inverse

In this section, we focus on the following problem. Let $\Omega \subset \mathbf{R}^n$ be an open set and assume that $f : \Omega \to \mathbf{R}^n$ is a homeomorphism and belongs to the Sobolev space $W^{1,p}(\Omega, \mathbf{R}^n)$, $p \geq 1$. Can we then conclude that the inverse mapping is also weakly differentiable, i.e. $f^{-1} \in W_{\mathrm{loc}}^{1,1}(f(\Omega), \Omega)$? In Chap. 1 we have seen the solution in the planar case and now we deal with higher dimensions. Let us first recall that in Example 1.1 we constructed a planar homeomorphism such that f is Lipschitz, but $f^{-1} \notin W_{\mathrm{loc}}^{1,1}$. Analogously, we can set $f(x) = [f_1(x_1), x_2, \ldots, x_n]$ (with f_1 as in Example 1.1) and we obtain a Lipschitz homeomorphism with $f^{-1} \notin W_{\mathrm{loc}}^{1,1}$.

It follows that the inverse of a Lipschitz homeomorphism may fail to be weakly differentiable. However it is possible to show that it is differentiable in an even weaker sense. Namely the derivative of the inverse map is not necessarily an integrable function but it is indeed a Radon measure.

Definition 5.1. We say that a real-valued function h has bounded variation on Ω, $h \in BV(\Omega)$, if $h \in L^1(\Omega)$ and $D_i h = \mu_i$ are signed Radon measures with finite total variation:

$$\int_\Omega h D_i \varphi \, dx = -\int_\Omega \varphi \, d\mu_i, \text{ for all } \varphi \in C_0^\infty(\Omega).$$

Then $g \in BV(\Omega, \mathbf{R}^n)$ means that each component function g_j, $j \in \{1, \ldots, n\}$, of g belongs to $BV(\Omega)$.

S. Hencl and P. Koskela, *Lectures on Mappings of Finite Distortion*, Lecture Notes in Mathematics 2096, DOI 10.1007/978-3-319-03173-6_5,
© Springer International Publishing Switzerland 2014

Let us formulate the statement analogous to the planar case Theorem 1.6 and further, the more general result, without the assumption that f has finite distortion.

Theorem 5.2. *Let $\Omega \subset \mathbf{R}^n$ be an open set. Suppose that $f \in W^{1,n-1}(\Omega, \mathbf{R}^n)$ is a homeomorphism of finite distortion. Then $f^{-1} \in W^{1,1}_{\mathrm{loc}}(f(\Omega), \mathbf{R}^n)$ and has finite distortion.*

Theorem 5.3. *Let $\Omega \subset \mathbf{R}^n$ be an open set. Suppose that $f \in W^{1,n-1}(\Omega, \mathbf{R}^n)$ is a homeomorphism. Then $f^{-1} \in BV_{\mathrm{loc}}(f(\Omega), \mathbf{R}^n)$.*

The following lemma will be crucial for our proofs. It basically tells us that $(n-1)$-dimensional null sets on spheres are mapped to $(n-1)$-dimensional null sets. Since the proof is quite technical, we omit it; we refer the interested reader to [20].

Lemma 5.4. *Let $f \in W^{1,n-1}(B(x, r_0), \mathbf{R}^n)$ be a homeomorphism. Then, for almost every $r \in (0, r_0)$, the mapping $f : S^{n-1}(x, r) \to \mathbf{R}^n$ satisfies the Lusin (N) condition, i.e.*

$$\mathcal{H}^{n-1}(f(A)) = 0 \text{ for every } A \subset S^{n-1}(x, r) \text{ such that } \mathcal{H}^{n-1}(A) = 0.$$

In this section we use the notation $\pi_r(x) = |x|$ for the radial projection and $\pi_S(x) = \frac{x}{|x|}$ for the projection to the unit sphere. The following coarea formula is the key ingredient in our proofs of Theorems 5.2 and 5.3.

Lemma 5.5. *Let $f \in W^{1,n-1}(\Omega, \mathbf{R}^n)$ be a homeomorphism. Set $h = \pi_S \circ f$ and let $E \subset \Omega$ be a measurable set. Then*

$$\int_{\partial B(0,1)} \mathcal{H}^1\big(\pi_r(\{x \in E : h(x) = z\})\big) \, d\mathcal{H}^{n-1}(z) \leq \int_E |\operatorname{adj} Dh| \, dx.$$

Proof. It is well known that Hausdorff measure does not increase under projections. Thus, for f Lipschitz, we can use the coarea formula Theorem A.38 to obtain

$$\int_{\partial B(0,1)} \mathcal{H}^1\big(\pi_r(\{x \in E : h(x) = z\})\big) \, d\mathcal{H}^{n-1}(z)$$

$$\leq \int_{\partial B(0,1)} \mathcal{H}^1\big(\{x \in E : h(x) = z\}\big) \, d\mathcal{H}^{n-1}(z) = \int_E |\operatorname{adj} Dh| \, dx.$$

In the general case, we proceed similarly to the proof of Theorem A.35. We cover the domain of f up to a set of measure zero by countably many sets of the type $\{f = f_j\}$ with f_j Lipschitz.

It remains to consider the case that $E = N$ with $|N| = 0$. For $z \in S^{n-1}(0, 1)$ we denote $S_z = \pi_S^{-1}(z)$. Suppose on the contrary that there is a set $P \subset S^{n-1}(0, 1)$ such that $\mathcal{H}^{n-1}(P) > 0$ and for every $z \in P$ we have $\mathcal{H}^1\big(\pi_r(f^{-1}(S_z) \cap E)\big) > 0$. Consider the set $A \subset (0, \infty) \times S^{n-1}(0, 1)$ defined by

$$[t, z] \in A \Leftrightarrow z \in P \text{ and } t \in \pi_r(f^{-1}(S_z) \cap E).$$

By the Fubini theorem we obtain

$$|A| = \int_P \mathcal{H}^1(\pi_r(f^{-1}(S_z) \cap E)) d\mathcal{H}^{n-1}(z) > 0.$$

Set $E_t = E \cap S^{n-1}(x, t)$. For almost every t we have $\mathcal{H}^{n-1}(E_t) = 0$ and therefore we obtain $\mathcal{H}^{n-1}(\pi_S \circ f(E_t)) = 0$ for almost every t by Lemma 5.4. Now the Fubini theorem implies

$$|A| = \int_0^\infty \mathcal{H}^{n-1}(\pi_S \circ f(E_t)) dt = 0$$

which gives us a contradiction. □

The following lemma shows that a Poincaré-type inequality holds for f^{-1}, which will give us the desired regularity of f^{-1}.

Lemma 5.6. *Let* $f \in W^{1,n-1}_{loc}(\Omega, \mathbf{R}^n)$ *be a homeomorphism. Then*

$$\int_B |f^{-1}(y) - f_B^{-1}| \, dy \le Cr_0 \int_{f^{-1}(B)} |\operatorname{adj} Df(x)| \, dx, \tag{5.1}$$

for each ball $B = B(y_0, r_0) \subset f(\Omega)$, *where* $C = C(n)$.

Proof. We fix $y' = f(x') \in B$ and for simplicity of notation (without loss of generality) we assume that $x' = 0$. Denote

$$h(x) = \frac{f(x) - y'}{|f(x) - y'|}.$$

If $y'' = f(x'') \in B$ and $[y'', y']$ is the line segment connecting y' and y'', then $f^{-1}([y'', y'])$ is a curve connecting x' and x'' and thus

$$|x'' - x'| \le \mathcal{H}^1(\pi_r \circ f^{-1}([y'', y'])). \tag{5.2}$$

We have

$$y \in [y'', y'] \implies \frac{y - y'}{|y - y'|} = \frac{y'' - y'}{|y'' - y'|}. \tag{5.3}$$

Hence, if $t = |y'' - y'|$, then

$$|f^{-1}(y'') - f^{-1}(y')| \le \mathcal{H}^1(\pi_r \circ f^{-1}([y'', y']))$$

$$\le \mathcal{H}^1\left(\pi_r\left(\{x \in f^{-1}(B) : h(x) = \tfrac{y''-y'}{t}\}\right)\right).$$

Given $t > 0$, using Lemma 5.5 for the mapping $f(x) - y'$ we estimate

$$\int_{B \cap \partial B(y',t)} |f^{-1}(y'') - f^{-1}(y')| \, d\mathcal{H}^{n-1}(y'')$$

$$\leq \int_{B \cap \partial B(y',t)} \mathcal{H}^1\left(\pi_r\left(\{x \in f^{-1}(B) : h(x) = \tfrac{y''-y'}{t}\}\right)\right) d\mathcal{H}^{n-1}(y'')$$

$$\leq t^{n-1} \int_{\partial B(0,1)} \mathcal{H}^1\left(\pi_r\left(\{x \in f^{-1}(B) : h(x) = z\}\right)\right) d\mathcal{H}^{n-1}(z) \qquad (5.4)$$

$$\leq t^{n-1} \int_{f^{-1}(B)} |\operatorname{adj} Dh(x)| \, dx$$

$$\leq C t^{n-1} \int_{f^{-1}(B)} \frac{|\operatorname{adj} Df(x)|}{|f(x) - f(x')|^{n-1}} \, dx,$$

where the last inequality follows using the chain rule, the inequality $|\operatorname{adj}(PQ)| \leq C |\operatorname{adj} P| |\operatorname{adj} Q|$ and the estimate

$$\left|\operatorname{adj} \nabla \frac{z - y'}{|z - y'|}\right| \leq \frac{C}{|z - y'|^{n-1}}.$$

Hence for $c = f_B^{-1}$ we have

$$|B| \, |f^{-1}(y') - c| \leq \int_B |f^{-1}(y'') - f^{-1}(y')| \, dy''$$

$$= \int_0^{2r_0} \left(\int_{B \cap \partial B(y',t)} |f^{-1}(y'') - f^{-1}(y')| \, d\mathcal{H}^{n-1}(y'')\right) dt$$

$$\leq C \int_0^{2r_0} t^{n-1} \left(\int_{f^{-1}(B)} \frac{|\operatorname{adj} Df(x)|}{|f(x) - f(x')|^{n-1}} \, dx\right) dt$$

$$\leq C r_0^n \int_{f^{-1}(B)} \frac{|\operatorname{adj} Df(x)|}{|f(x) - f(x')|^{n-1}} \, dx.$$

Integrating with respect to y' and then using the Fubini theorem we obtain

$$\int_B |f^{-1}(y') - c| \, dy' \leq C \int_{f^{-1}(B)} |\operatorname{adj} Df(x)| \left(\int_B \frac{dy'}{|f(x) - y'|^{n-1}}\right) dx$$

$$\leq C r_0 \int_{f^{-1}(B)} |\operatorname{adj} Df(x)| \, dx. \qquad \square$$

Proof (of Theorem 5.3). It follows from Lemma 5.6 that there is a locally finite Radon measure μ such that f^{-1} satisfies

$$\inf_{c \in \mathbf{R}} \int_B |f^{-1}(y) - c| \, dy \leq r_0 \mu(B)$$

for every ball $B \subset f(\Omega)$. Thus $f^{-1} \in BV_{\text{loc}}$ (see Theorem A.21).   □

We will introduce some notation needed in the sequel. We write $\mathbb{H}_i$ for the i-th coordinate hyperplane

$$\mathbb{H}_i = \{x \in \mathbf{R}^n : x_i = 0\}$$

and denote by π_i the orthogonal projection to $\mathbb{H}_i$, so that

$$\pi_i(x) = x - x_i \mathbf{e}_i, \qquad x \in \mathbf{R}^n.$$

By π^j we denote the projection to the j-th coordinate: $\pi^j(x) = x_j$.

Theorem 5.7. *Let $\Omega \subset \mathbf{R}^n$ be an open set and let $f \in W_{\text{loc}}^{1,n-1}(\Omega, \mathbf{R}^n)$ be a homeomorphism. Then, for each measurable set $E \subset \Omega$, we have*

$$\int_E |\operatorname{adj} D(\pi_i \circ f)| = \int_{\pi_i(\mathbf{R}^n)} \mathscr{H}^1(E \cap (\pi_i \circ f)^{-1}(y)) \, dy.$$

Proof. As in the proof of Lemma 5.5, we can rely on the Lipschitz coarea formula, Theorem A.38, and then restrict our attention to the case when E is Lebesgue null. We only need to show that the set of all points $y \in \pi_i(\mathbf{R}^n)$ such that $\mathscr{H}^1(E \cap (\pi_i \circ f)^{-1}(y)) > 0$ is of measure zero. Since we already know that f^{-1} is of bounded variation, it follows that for almost every y the preimage $(\pi_i \circ f)^{-1}(y)$ is a rectifiable curve (see Theorem A.22). Hence, if this is of positive one-dimensional measure, there exists a one-dimensional projection of this set which is also of positive one-dimensional measure. Now we can obtain a contradiction exactly as in the proof of Lemma 5.5.   □

We show that each bi-Sobolev mapping with $f \in W^{1,n-1}$ must be a mapping of finite distortion.

Theorem 5.8. *Let $\Omega \subset \mathbf{R}^n$ be an open set and let $f \in W_{\text{loc}}^{1,n-1}(\Omega, \mathbf{R}^n)$ be a homeomorphism such that $f^{-1} \in W_{\text{loc}}^{1,1}(f(\Omega), \mathbf{R}^n)$. Then f^{-1} is a mapping of finite distortion.*

Proof. Suppose that f^{-1} is not a mapping of finite distortion. Then we can find a set $\tilde{A} \subset f(\Omega)$ such that $|\tilde{A}| > 0$ and for every $y \in \tilde{A}$ we have $J_{f^{-1}}(y) = 0$ and $|Df^{-1}(y)| > 0$. Since f^{-1} is of the class $W_{\text{loc}}^{1,1}$, we may assume without loss of generality that f^{-1} is absolutely continuous on all lines parallel to coordinate axes that intersect $\tilde{A}$ (see Theorem A.15) and that f^{-1} has classical partial derivatives at every point of $\tilde{A}$, because absolutely continuous functions of one variable are differentiable a.e.

We claim that we can find a Borel set $A \subset \tilde{A}$ such that $|A| > 0$ and $|f^{-1}(A)| = 0$. We know that the Lusin (N) condition and hence also the Area formula holds on a set of full measure, see Corollary A.36 (c), and thus we can find a Borel set $A \subset \tilde{A}$ such that $|A| > 0$ and the Lusin (N) condition holds for f^{-1} on A. By the Area formula for f^{-1} we now obtain

$$|f^{-1}(A)| = \int_{\Omega} \chi_{f^{-1}(A)}(x)\, dx = \int_{f(\Omega)} \chi_A(y)|J_{f^{-1}}(y)|\, dy = 0.$$

Clearly, there is $i \in \{1 \ldots, n\}$ such that the subset of A where $\frac{\partial f^{-1}(y)}{\partial y_i} \neq 0$ has positive measure. Without loss of generality we will assume that $\frac{\partial f^{-1}(y)}{\partial y_i} \neq 0$ for every $y \in A$. Set $E := f^{-1}(A)$ and recall that $|E| = 0$. Using Theorem 5.7 we obtain

$$\int_{\mathbb{H}_i} \mathscr{H}^1\big(\pi^j(\{x \in E : \pi_i \circ f(x) = z\})\big)dz = 0, \qquad (5.5)$$

for each $j \in \{1, \ldots, n\}$. By the Fubini theorem,

$$\int_{\mathbb{H}_i} \mathscr{H}^1(A \cap \pi_i^{-1}(z))\, dz = |A| > 0.$$

Therefore there exists $z \in \mathbb{H}_i$ with

$$\mathscr{H}^1\big(\pi^j(E \cap f^{-1}(\pi_i^{-1}(z)))\big) = \mathscr{H}^1\big(\pi^j(\{x \in E : \pi_i \circ f(x) = z\})\big) = 0,$$

and

$$\mathscr{H}^1(A \cap \pi_i^{-1}(z)) > 0.$$

Clearly

$$0 < \int_{A \cap \pi_i^{-1}(z)} \left|\frac{\partial f^{-1}}{\partial y_i}(y)\right| d\mathscr{H}^1(y)$$

and therefore we can find j such that for $h = \pi^j \circ f^{-1}$ we have

$$0 < \int_{A \cap \pi_i^{-1}(z)} \left|\frac{\partial h}{\partial y_i}(y)\right| d\mathscr{H}^1(y).$$

The mapping

$$t \mapsto h(z + t e_i)$$

is absolutely continuous and therefore satisfies the Lusin (N) condition. Hence we may apply the one-dimensional Area formula, Theorem A.35, to obtain

$$0 < \int_{A \cap \pi_i^{-1}(z)} \left| \frac{\partial h}{\partial y_i}(y) \right| d\mathcal{H}^1(y)$$

$$= \int_{\mathbf{R}} N(h, A \cap \pi_i^{-1}(z), x) \, dx$$

$$= \int_{\pi^j(E \cap f^{-1}(\pi_i^{-1}(z)))} N(h, A \cap \pi_i^{-1}(z), x) \, dx$$

$$= 0 ,$$

which is a contradiction. ☐

Proof (of Theorem 5.2). We claim that there is a function $g \in L^1_{\mathrm{loc}}(f(\Omega))$ such that

$$\int_{f^{-1}(B)} |\operatorname{adj} Df| \le \int_B g. \tag{5.6}$$

This and Lemma 5.6 imply that the pair f, g satisfies a 1-Poincaré inequality in $f(\Omega)$. From Theorem A.20 we then deduce that $f^{-1} \in W^{1,1}_{\mathrm{loc}}(f(\Omega), \mathbf{R}^n)$.

By Corollary A.36 (b) there is a set $\Omega' \subset \Omega$ of full measure such that the Area formula (A.20) holds for f on Ω'. We define a function $g: f(\Omega) \to \mathbf{R}$ by setting

$$g(f(x)) = \begin{cases} \frac{|\operatorname{adj} Df(x)|}{J_f(x)} & \text{if } x \in \Omega' \text{ and } J_f(x) > 0, \\ 0 & \text{otherwise.} \end{cases}$$

Since f is a mapping of finite distortion, we have

$$|\operatorname{adj} Df(x)| = g(f(x)) \, J_f(x) \text{ a.e. in } \Omega.$$

Hence for every Borel set $A \subset f(\Omega)$

$$\int_{f^{-1}(A)} |\operatorname{adj} Df(x)| \, dx = \int_{f^{-1}(A) \cap \Omega'} g(f(x)) \, J_f(x) \, dx$$

$$\le \int_A g(y) \, dy. \tag{5.7}$$

For $A = B$ this gives (5.6) and by arbitrariness of A it also implies $g \in L^1_{\mathrm{loc}}$.
From Theorem 5.8 we now obtain that f^{-1} has finite distortion. ☐

Analogously to Theorem 1.7 we obtain $W^{1,n}$-regularity of the inverse if K_f^{n-1} is integrable.

Theorem 5.9. *Let $\Omega \subset \mathbf{R}^n$ be an open set. Suppose that $f \in W^{1,1}(\Omega, \mathbf{R}^n)$ is a homeomorphism of finite distortion with $K_f \in L^{n-1}(\Omega)$. Then $f^{-1} \in W^{1,n}_{\mathrm{loc}}(f(\Omega), \mathbf{R}^n)$ and f^{-1} is a mapping of finite distortion.*

Proof. By the distortion inequality and Hölder's inequality we have

$$\int_A |Df(x)|^{n-1}\,dx \le \int_A K_f(x)^{\frac{n-1}{n}} J_f(x)^{\frac{n-1}{n}}\,dx \le \|K_f\|_{L^{n-1}(A)}^{\frac{n-1}{n}} \|J_f\|_{L^1(A)}^{\frac{n-1}{n}}$$

and thus $f \in W^{1,n-1}_{\mathrm{loc}}(\Omega)$. From Theorem 5.2 we already know that $f^{-1} \in W^{1,1}_{\mathrm{loc}}$ and that f^{-1} is a mapping of finite distortion. Therefore it is enough to prove that $\int_{f(\Omega)} |Df^{-1}|^n$ is finite.

We consider the integral over the set

$$A = \{y \in f(\Omega) : \ f^{-1} \text{ is approximatively differentiable at } y \text{ and } J_{f^{-1}}(y) > 0\}.$$

We know that f^{-1} is approximatively differentiable a.e. (see Theorem A.31) and that f^{-1} is a mapping of finite distortion by Theorem 5.8. It follows that $|Df^{-1}|^n$ vanishes a.e. on $f(\Omega) \setminus A$ and hence

$$\int_{f(\Omega)} |Df^{-1}(y)|^n\,dy = \int_A |Df^{-1}(y)|^n\,dy .$$

From Lemma A.33 we obtain that f is approximatively differentiable at $x = f^{-1}(y)$ for a.e. every $y \in A$ with

$$Df^{-1}(f(x))Df(x) = I \text{ and } J_f^{-1}(f(x))J_f(x) = 1 .$$

We pick a Borel set $\tilde{A} \subset A$ with $|\tilde{A}| = |A|$ and so that the above holds on $\tilde{A}$. We conclude using the Area formula, Corollary A.36 (a), and the identity $M \operatorname{adj} M = I \det M$ that

$$\int_{f(\Omega)} |Df^{-1}(y)|^n\,dy = \int_{\tilde{A}} \frac{|Df^{-1}(y)|^n}{J_{f^{-1}}(y)} J_{f^{-1}}(y)\,dy \le \int_{f^{-1}(\tilde{A})} \frac{|Df^{-1}(f(x))|^n}{J_{f^{-1}}(f(x))}\,dx$$

$$= \int_{f^{-1}(\tilde{A})} |(Df(x))^{-1}|^n J_f(x)\,dx = \int_{f^{-1}(\tilde{A})} \frac{|\operatorname{adj} Df(x)|^n}{J_f(x)^{n-1}}\,dx$$

$$\le \int_{f^{-1}(\tilde{A})} \frac{|Df(x)|^{(n-1)n}}{J_f(x)^{n-1}}\,dx \le \int_\Omega K_f(x)^{n-1}\,dx.$$

$$(5.8)$$

$\square$

The following counterexample shows that Theorems 5.2 and 5.3 are sharp in the sense that the crucial regularity assumption $|Df| \in L^{n-1}$ cannot be essentially relaxed.

Example 5.10. Let $0 < \varepsilon < 1$ and $n \geq 3$. There is a homeomorphism f : $(-1, 1)^n \rightarrow \mathbf{R}^n$ such that $f \in W^{1, n-1-\varepsilon}((-1, 1)^n, \mathbf{R}^n)$, f^{-1} is continuously differentiable at every point of

$$f((-1, 1)^n) \setminus \{[0, \dots, 0, t] \in \mathbf{R}^n : t \in \mathbf{R}\}$$

and $|\nabla f^{-1}| \notin L^1_{loc}(f((-1, 1)^n))$, where ∇f^{-1} denotes the pointwise differential of f^{-1}. Consequently, $f^{-1} \notin BV_{loc}(f(\Omega), \mathbf{R}^n)$.

Proof. We write $\mathbf{e}_i$ for the i-th unit vector in $\mathbf{R}^n$, i.e. the vector with 1 on the i-th place and 0 everywhere else. Given $x = [x_1, \dots, x_n] \in \mathbf{R}^n$ we denote $\tilde{x} = [x_1, \dots, x_{n-1}] \in \mathbf{R}^{n-1}$ and $|\tilde{x}| = \sqrt{x_1^2 + \dots + x_{n-1}^2}$.

Let $\alpha = \frac{\varepsilon}{n^2}$, $\beta = 1 + \frac{\varepsilon}{(n-1)}$ and set

$$f(x) = \sum_{i=1}^{n-1} \mathbf{e}_i x_i |\tilde{x}|^{\alpha-1} + \mathbf{e}_n \left(x_n + |\tilde{x}| \sin(|\tilde{x}|^{-\beta}) \right)$$

if $|\tilde{x}| > 0$ and $f(x) = \mathbf{e}_n x_n$ if $|\tilde{x}| = 0$. Our mapping f is clearly continuous and it is easy to check that f is a one-to-one map since

$$x_i |\tilde{x}|^{\alpha-1} = z_i |\tilde{z}|^{\alpha-1} \text{ for every } i \in \{1, \dots, n-1\}$$
$$\Rightarrow x_i = z_i \text{ for every } i \in \{1, \dots, n-1\}.$$

Therefore f is a homeomorphism.

A direct computation shows that the absolute values of the partial derivatives of $f_i, i \in \{1, \dots, n-1\}$, are smaller than $C|\tilde{x}|^{\alpha-1}$ and therefore integrable with the exponent $p = n - 1 - \varepsilon$. Moreover,

$$\frac{\partial f_n(x)}{\partial x_1} = x_1 |\tilde{x}|^{-1} \sin(|\tilde{x}|^{-\beta}) - |\tilde{x}| \beta \frac{x_1}{|\tilde{x}|^{\beta+2}} \cos(|\tilde{x}|^{-\beta}).$$

The first term is clearly integrable with the exponent $n - 1 - \varepsilon$. For this degree of integrability of the second one, we need

$$(\beta + 2 - 1 - 1)(n - 1 - \varepsilon) < n - 1$$

which is guaranteed by our choice of β. Analogously $\frac{\partial f_n}{\partial x_i}$ is integrable for $i = 2, \dots, n - 1$. Finally, $\frac{\partial f_n}{\partial x_n}$ is bounded. Since f is C^1-smooth outside the segment $\{[0, \dots, 0, t] : t \in (-1, 1)\}$ and $|\nabla f| \in L^{n-1-\varepsilon}((-1, 1)^n)$ it is easy to see that $f \in W^{1, n-1-\varepsilon}((-1, 1)^n, \mathbf{R}^n)$.

The inverse of f is given by

$$f^{-1}(y) = \sum_{i=1}^{n-1} \mathbf{e}_i \, y_i \, |\tilde{y}|^{\frac{1}{\alpha}-1} + \mathbf{e}_n \left(y_n - |\tilde{y}|^{\frac{1}{\alpha}} \sin(|\tilde{y}|^{-\frac{\beta}{\alpha}}) \right)$$

if $|\tilde{y}| > 0$ and $f^{-1}(y) = \mathbf{e}_n y_n$ if $|\tilde{y}| = 0$. The differential of f^{-1} is clearly continuous outside the segment $\{[0,\cdots,0,t] : t \in \mathbf{R}\}$. Computations as above show us that for $|\nabla f^{-1}| \in L^1_{\text{loc}}$ we need that

$$\frac{\beta}{\alpha} + 2 - \frac{1}{\alpha} - 1 < n - 1.$$

This is not satisfied and therefore $|\nabla f^{-1}|$ is not locally integrable.   □

Remark 5.11. (a) The planar case studied in our first chapter was treated by Hencl and Koskela [49].

(b) The BV-regularity of the inverse was first studied by Hencl et al. [51]. The full proofs of Theorems 5.2 and 5.3 and especially Lemma 5.4 can be found in Csörnyei et al. [20].

(c) Example 5.10 is based on [51, Example 3.1] (also see the examples in [50] and [46]). For more information see [25] and [24].

(d) By Theorem 5.8 we know that each bi-Sobolev mapping f with $f \in W^{1,n-1}$ is a mapping of finite distortion. This is not true without the assumption $f \in W^{1,n-1}$, but each bi-Sobolev mapping satisfies $J_f(x) = 0 \Rightarrow |\operatorname{adj} Df(x)| = 0$ for every $x \in \Omega \setminus N$, where $|N| = 0$. These results were established in [55].

(e) Tengvall has very recently shown in [119] that a homeomorphism $f \in W^{1,n-1}$ with finite distortion and $K_f \in L^{n-1}$ is differentiable a.e. In fact, he proves this for discrete and open mappings in this class.

(f) By Theorem 5.2 we know that f^{-1} is a mapping of finite distortion if $f \in W^{1,n-1}$ is a homeomorphic mapping with finite distortion. What then about the degree of integrability of $K_{f^{-1}}$? For simplicity let us assume that $K_f \in L^{n-1}$. Then $f^{-1} \in W^{1,n}_{\text{loc}}$, it is differentiable a.e. and satisfies the Lusin condition (N). Hence we can make computations in the spirit of (5.8) to see that

$$\int_{f(\Omega)} K_{f^{-1}}(y)\, dy = \int_{f^{-1}(A)} \frac{|Df^{-1}(f(x))|^n}{J_{f^{-1}}(f(x))} J_f(x)\, dx$$

$$= \int_{f^{-1}(A)} |(Df(x))^{-1}|^n J_f^2(x)\, dx$$

$$= \int_{f^{-1}(A)} \frac{|\operatorname{adj} Df(x)|^n}{J_f(x)^{n-2}}\, dx \le \int_{\Omega} K_f(x)^{n-1} J_f(x)\, dx.$$

Thus integrability properties of $K_{f^{-1}}$ depend on the integrability degrees of K_f and J_f. For concrete results in this direction see e.g. [37,38,49,55,98,104].

5.2 Regularity of the Composition

It is well known that the composition of a quasiconformal mapping $f : \Omega \to \mathbf{R}^n$ and a function $u \in W^{1,n}_{loc}(f(\Omega), \mathbf{R}^n)$ satisfies $u \circ f \in W^{1,n}_{loc}(\Omega, \mathbf{R}^n)$. The aim of this section is to generalize this result and to characterize those homeomorphisms f for which the composition operator $u \circ f$ maps one Sobolev space to another, possibly different Sobolev space.

Definition 5.12. Let Ω_1, Ω_2 be open subsets of $\mathbf{R}^n$, $1 \le p \le q < \infty$ and let f be a mapping from Ω_1 to Ω_2. We say that the operator T_f defined by

$$(T_f u)(x) := u(f(x)) \text{ for } x \in \Omega_1,$$

is continuous from $W^{1,q}_{loc}(\Omega_2)$ (respectively $W^{1,q}_{loc}(\Omega_2) \cap C(\Omega_2)$) into $W^{1,p}_{loc}(\Omega_1)$, if $T_f u \in W^{1,p}_{loc}(\Omega_1)$ for all functions $u \in W^{1,q}_{loc}(\Omega_2)$ (resp. $W^{1,q}_{loc}(\Omega_2) \cap C(\Omega_2)$). Moreover we require that there is a constant C independent of u such that

$$\|DT_f u\|_{L^p(\Omega_1)} \le C \|Du\|_{L^q(\Omega_2)}. \tag{5.9}$$

For $q \in [1, \infty)$ and mappings of finite distortion f we define the function K_q which generalizes the usual distortion by setting

$$K_q(x) = \begin{cases} \frac{|Df(x)|^q}{|J_f(x)|} & \text{if } J_f(x) > 0, \\ 0 & \text{otherwise.} \end{cases}$$

Theorem 5.13. Let Ω_1, Ω_2 be open subsets of $\mathbf{R}^n$, $1 \le p \le q < \infty$, and let $f \in W^{1,1}_{loc}(\Omega_1, \Omega_2)$ be a homeomorphism of finite distortion satisfying

$$K_q(x) \in L^{\frac{p}{q-p}}(\Omega_1). \tag{5.10}$$

Then the operator T_f is continuous from $W^{1,q}_{loc}(\Omega_2)$ into $W^{1,p}_{loc}(\Omega_1)$ if $1 \le q \le n$ and the operator is continuous from $W^{1,q}_{loc}(\Omega_2) \cap C(\Omega_2)$ to $W^{1,p}_{loc}(\Omega_1)$ if $n < q < \infty$.
   Moreover we have

$$D(u \circ f)(x) = Du(f(x))Df(x) \text{ for a.e. } x \in \Omega_1 \tag{5.11}$$

if we use the convention that $Du(f(x)) \cdot 0 = 0$ even if Du does not exist or it equals infinity at $f(x)$.

Let us note that the assumption that u is continuous for $q > n$ is crucial. Otherwise the composition $u \circ f$ may even fail to be measurable if f does not satisfy the Lusin (N^{-1}) condition. For $q > n$ we can always choose a continuous representative of $u \in W^{1,p}$ and so this is not a big inconvenience. On the other hand, the statement is valid for every representative of f for $q \le n$ because of the validity of the Lusin (N^{-1}) condition.

Theorem 5.14. *Suppose that* $f \in W^{1,1}_{loc}(\Omega, \mathbf{R}^n)$ *is a homeomorphism with finite distortion satisfying*

$$K_q \in L^{\frac{1}{q-1}}_{loc}(\Omega)$$

for some $q \in [1, n]$. *Then* f *satisfies the Lusin* (N^{-1}) *condition.*

Proof. Let us denote $Z = \{x : |Df(x)| = 0\}$. For each ball $B \subset \Omega$ we have

$$\int_B K_n^{\frac{1}{n-1}}(x)\, dx = \int_{B \setminus Z}\Big(\frac{|Df(x)|^n}{J_f(x)}\Big)^{\frac{1}{n-1}}\, dx = \int_{B \setminus Z}\Big(\frac{|Df(x)|^q}{J_f(x)}\Big)^{\frac{1}{n-1}}|Df(x)|^{\frac{n-q}{n-1}}\, dx.$$

By the Hölder inequality and $\frac{p}{q-p} \ge \frac{1}{q-1}$ we obtain

$$\int_B K_n^{\frac{1}{n-1}}(x)\, dx \le \|K_q^{\frac{1}{q-1}}\|_{L^1(B)}^{\frac{q-1}{n-1}} \|Df\|_{L^1(B)}^{\frac{n-q}{n-1}} < \infty.$$

It follows that the usual distortion function satisfies $K_n \in L^{\frac{1}{n-1}}_{loc}(\Omega)$ and the result follows by Theorem 4.13. $\square$

The second ingredient is the following approximation by Lipschitz functions that we know from the construction in Lemma A.25.

Lemma 5.15. *Let* $1 \le q < \infty$ *and let* $u \in W^{1,q}(B(x_0, 3r))$. *There is a sequence of functions* u_k *with Lipschitz constant* $C \cdot k$ *and sequence of measurable sets* F_k *such that* $F_k \subset \{u = u_k\}$, $F_k \subset F_{k+1}$

$$\lim_{k \to \infty} \mathcal{L}_n(B(x_0, r) \setminus F_k) = 0 \text{ and } u_k \overset{k \to \infty}{\to} u \text{ in } W^{1,q}(B).$$

If $q > n$ *then* u_k *converges to* $\tilde{u}$ *uniformly, where* $\tilde{u}$ *is the continuous representative of* u.

Lemma 5.16. *Let* $\Omega \subset \mathbf{R}^n$ *be an open set and* $p \in [1, \infty)$. *Suppose that* $u : \mathbf{R}^n \to \mathbf{R}$ *is a Lipschitz function,* $f \in W^{1,p}(\Omega, \mathbf{R}^n)$ *has finite distortion and that* $u \circ f \in L^p$. *Then* $u \circ f \in W^{1,p}(\Omega)$ *and we have*

$$D(u \circ f)(x) = \nabla u(f(x)) \cdot Df(x) \text{ for a.e. } x \in \Omega, \tag{5.12}$$

where we define $\nabla u(f(x)) \cdot Df(x) := 0$ *if* $\nabla u(f(x))$ *does not exist or equals infinity and* $Df(x) = 0$.

Proof. We may assume that the component functions f_i are absolutely continuous on almost all line segments in Ω parallel to coordinate axes for all $1 \le i \le n$. Then clearly $u \circ f$ is absolutely continuous on the same lines because u is Lipschitz and hence $\frac{\partial u \circ f}{\partial x_i}$ exists a.e. Moreover,

$$\frac{\partial u \circ f}{\partial x_i}(x) = \sum_{j=1}^{n} \frac{\partial u}{\partial y_j}(f(x)) \frac{\partial f_j}{\partial x_i}(x), \qquad (5.13)$$

holds at all $x \in \Omega$ such that $\nabla u(f(x))$ and $\frac{\partial f_j}{\partial x_i}(x)$ both exist. Note that if $\frac{\partial f_j}{\partial x_i}(x) = 0$ for all $1 \leq j \leq n$ then $\frac{\partial u \circ f}{\partial x_i}(x) = 0$ because

$$\frac{\left| u(f(x + he_i)) - u(f(x)) \right|}{h} \leq C \frac{\| f(x + he_i) - f(x) \|}{h} \stackrel{h \to 0+}{\to} 0.$$

Hence, if we use the convention $\frac{\partial u}{\partial y_j}(f(x)) \cdot 0 = 0$, then (5.13) holds also in this case. Now we show that the set

$$N = \left\{ x \in \Omega : \nabla u(f(x)) \text{ does not exist and } \frac{\partial f_j}{\partial x_i}(x) \neq 0 \text{ for some } 1 \leq j \leq n \right\}$$

has zero measure. Let Z be a zero measure Borel measurable set such that u is differentiable on $\mathbf{R}^n \setminus Z$. By the Area formula (Theorem A.35) we have

$$\int_{f^{-1}(Z)} |J_f| \leq \mathscr{L}_n(Z) = 0.$$

Therefore $J_f = 0$ a.e. on $f^{-1}(Z)$ and hence $Df = 0$ a.e. on $f^{-1}(Z)$, because f has finite distortion. Since $N \subset f^{-1}(Z)$ it clearly follows from the definition of N that $\mathscr{L}_n(N) = 0$. Thus (5.13) holds a.e. on Ω with our convention. At all such x, we may conclude that

$$\left| \frac{\partial u \circ f}{\partial x_i}(x) \right|^p \leq C \sup_{1 \leq i \leq n} \left| \frac{\partial f_j}{\partial x_i}(x) \right|^p. \qquad (5.14)$$

Since $u \circ f$ satisfies the ACL-condition, we obtain by integration that $u \circ f \in W^{1,p}$.
□

Proof (of Theorem 5.13). Let $u \in W^{1,q}_{loc}(\Omega_2)$ be arbitrary and let $x_0 \in \Omega_1$. Fix a ball B and $r > 0$ such that $3B \subset\subset \Omega_2$ and $f(B(x_0, r)) \subset B$. We have to show that $u \circ f \in W^{1,1}(B(x_0, r))$ and $|D(u \circ f)|$ in $L^p(B(x_0, r))$.

By applying Lemma 5.15 we find a sequence of Lipschitz functions u_k and a sequence of measurable sets $F_k \subset B$ such that

$$u_k = u \text{ on } F_k, \ F_k \subset F_{k+1} \text{ and } \lim_{k \to \infty} \mathscr{L}_n(F_k) = \mathscr{L}_n(B).$$

Set $g_j = u_j \circ f$ for each $j \in \mathbf{N}$. Since u_j is Lipschitz, we obtain from Lemma 5.16 that $g_j \in W^{1,p}(B(x_0, r))$.

We show that Dg_j form a Cauchy sequence in $L^p(B(x_0, r), \mathbf{R}^n)$. Let v be a Lipschitz function. From Lemma 5.16 and the definition of q-distortion we have

$$\int_{B(x_0,r)} \left| D\big(v(f(x))\big) \right|^p dx \leq \int_{B(x_0,r)} \left| \nabla v(f(x)) \right|^p |Df(x)|^p dx$$

$$\leq \int_{B(x_0,r)} \left| \nabla v(f(x)) \right|^p |J_f(x)|^{\frac{p}{q}} \big(K_q(x)\big)^{\frac{p}{q}} dx.$$

Now, if we use the Hölder inequality and the Area formula (Theorem A.35) (it follows from Lemma 5.16 that it does not matter how we redefine ∇v in the set where ∇v does not exist, and so we may assume that ∇v is a Borel measurable function), we obtain

$$\int_{B(x_0,r)} \left| D\big(v(f(x))\big) \right|^p dx \leq \left(\int_{B(x_0,r)} |\nabla v(f(x))|^q |J_f(x)| dx \right)^{\frac{p}{q}} \left(\int_{B(x_0,r)} (K_q(x))^{\frac{p}{q}\frac{q}{q-p}} dx \right)^{\frac{q-p}{q}}$$

$$\leq \left(\int_{f(B(x_0,r))} (\nabla v(y))^q dy \right)^{\frac{p}{q}} \left(\int_{B(x_0,r)} (K_q(x))^{\frac{p}{q-p}} dx \right)^{\frac{q-p}{q}}$$

$$\leq \| \nabla v \|^p_{L^q(f(B(x_0,r)))} \| K_q \|^{\frac{p}{q}}_{L^{\frac{p}{q-p}}(B(x_0,r))}, \tag{5.15}$$

where the last norm is L^∞-norm in the case $p = q$. Let us note that this is the key inequality of the whole proof. This is the exact reason why our integrability condition on K_q was needed.

If we apply this estimate to the function $v = u_j - u_k$, we easily get that $D(u_j \circ f) = Dg_j$ form a Cauchy sequence in $L^p(B(x_0, r), \mathbf{R}^n)$. Hence there exists a mapping $h \in L^p(B(x_0, r), \mathbf{R}^n)$ such that

$$Dg_j \xrightarrow{j \to \infty} h \text{ in } L^p(B(x_0, r), \mathbf{R}^n). \tag{5.16}$$

Suppose that $q \leq n$. By Theorem 5.14 our mapping f satisfies the Lusin (N^{-1}) condition, and hence f^{-1} maps sets of measure zero to sets of measure zero. Since $\mathscr{L}_n(B \setminus F_j)$ converges to zero, we obtain that sets $A_j := B(x_0, r) \cap f^{-1}(F_j)$ satisfy

$$\lim_{j \to \infty} \mathscr{L}_n(A_j) = \mathscr{L}_n\left(B(x_0, r) \cap f^{-1}\big(\bigcup_{j=1}^{\infty} F_j\big) \right) = \mathscr{L}_n\big(B(x_0, r)\big).$$

Hence we can find j_0 such that $\mathscr{L}_n(A_{j_0}) \geq \frac{1}{2}\mathscr{L}_n\big(B(x_0, r)\big)$. From the definition of g_j we have $g_j(x) = u \circ f(x)$ for all $x \in A_{j_0}$, and hence $g_j(x) - g_i(x) = 0$ on A_{j_0} for all $i, j \geq j_0$. Write $g = g_i - g_j$. It follows from the Poincaré inequality Theorem A.17, $g_{A_{j_0}} = 0$ and $\mathscr{L}_n(A_{j_0}) \geq \frac{1}{2}\mathscr{L}_n\big(B(x_0, r)\big)$ that

$$\int_{B(x_0,r)} |g_i - g_j| = \int_{B(x_0,r)} |g| = \int_{B(x_0,r)} |g(x) - g_{A_{j_0}}| dx$$

$$\leq \int_{B(x_0,r)} |g(x) - g_{B(x_0,r)}| dx + \mathscr{L}_n\big(B(x_0,r)\big)|g_{A_{j_0}} - g_{B(x_0,r)}|$$

$$\leq C(n)r \int_{B(x_0,r)} |Dg| + \frac{\mathscr{L}_n\big(B(x_0,r)\big)}{\mathscr{L}_n(A_{j_0})} \int_{A_{j_0}} |g(x) - g_{B(x_0,r)}| dx$$

$$\leq C(n,r) \int_{B(x_0,r)} |Dg| = C(n,r) \int_{B(x_0,r)} |Dg_j - Dg_i|.$$

Since $\{Dg_j\}$ is a Cauchy sequence in $L^1\big(B(x_0,r), \mathbf{R}^n\big)$ we obtain that $\{g_j\}$ is a Cauchy sequence in $L^1\big(B(x_0,r)\big)$. Moreover, the values of g_j converge to $u \circ f$ at points of $\bigcup_{j=1}^{\infty} A_j$, i.e. almost everywhere, and thus $g_j \to u \circ f$ in $L^1\big(B(x_0,r)\big)$. If $q > n$, then $u_j \to u$ uniformly by Lemma 5.15 and hence $u_j \circ f \to u \circ f$ uniformly which implies $g_j \to u \circ f$ in $L^1\big(B(x_0,r)\big)$.

The definition of the weak derivative gives us

$$\int_{B(x_0,r)} Dg_j(x)\varphi(x) = -\int_{B(x_0,r)} g_j(x)\nabla\varphi(x)$$

for each $\varphi \in C_C^{\infty}(B(x_0,r))$. Since $Dg_j \to h$ in $L^p\big(B(x_0,r), \mathbf{R}^n\big)$ and $g_j \to u \circ f$ in $L^1\big(B(x_0,r)\big)$, by passing j to infinity we get

$$\int_{B(x_0,r)} h(x)\varphi(x) = -\int_{B(x_0,r)} u \circ f(x)\nabla\varphi(x). \qquad (5.17)$$

This means that $h \in L^p\big(B(x_0,r), \mathbf{R}^n\big)$ is the weak gradient of $u \circ f$ on $B(x_0,r)$ and therefore $u \circ f \in W_{\text{loc}}^{1,p}(\Omega_1)$.

Now we want to show that T_f is continuous on the whole of Ω_1. For all balls $B(x_0,r)$ such that $f\big(B(x_0,r)\big) \subset B$ for some ball B with $3B \subset \Omega_2$, we have using (5.15)

$$\int_{B(x_0,r)} |D(u_k \circ f)|^p dx \leq \|\nabla u_k\|_{L^q(f(B(x_0,r)))}^p \|K_q\|_{L^{\frac{p}{q-p}}(B(x_0,r))}^{\frac{p}{q}}.$$

Since ∇u_k converges to Du in L^q and $D(u_k \circ f)$ converges to $D(u \circ f)$ in L^p, we obtain

$$\int_{B(x_0,r)} |D(u \circ f)|^p dx \leq \|Du\|_{L^q(f(B(x_0,r)))}^p \|K_q\|_{L^{\frac{p}{q-p}}(B(x_0,r))}^{\frac{p}{q}}. \qquad (5.18)$$

By applying Vitali's covering Theorem A.1 and our assumption $p \leq q$, we easily deduce the validity of the same inequality over Ω_1.

It remains to show the familiar formula (5.11). If $q \leq n$, then f satisfies the Lusin (N^{-1}) condition by Theorem 5.14 and hence $\nabla u_k(f(x))$ is well defined a.e. in Ω_1. We may assume without loss of generality that $\nabla u_k(x) \to Du(x)$ on $\Omega_2 \setminus N$, where N is a Borel set of measure zero. It easily follows that

$$\nabla u_k(f(x))Df(x) \overset{k\to\infty}{\to} Du(f(x))Df(x) \tag{5.19}$$

on $\Omega_1 \setminus f^{-1}(N)$, i.e. almost everywhere. From (5.16) and (5.17) we know that $D(u_k \circ f) = \nabla u_k \circ f \cdot Df$ converges to $D(u \circ f)$. It easily follows that

$$D(u \circ f)(x) = Du(f(x))Df(x) \text{ for a.e. } x \in \Omega_1. \tag{5.20}$$

If $q > n$ and the Lusin (N^{-1}) condition fails, then (5.20) holds again on $\Omega_1 \setminus f^{-1}(N)$, but $f^{-1}(N)$ may have positive measure. By the Area formula (Theorem A.35) we obtain that $J_f = 0$ on $f^{-1}(N)$ and hence $Df = 0$ a.e. on $f^{-1}(N)$ since f has finite distortion. By Lemma (5.16) we have $D(u_k \circ f) = 0$ a.e. on $f^{-1}(N)$ and hence

$$D(u \circ f) = 0 \text{ a.e. on } f^{-1}(N).$$

If we use the convention $Du(f(x)) \cdot 0 = 0$ (including the case when Du does not exist or is infinity at $f(x)$) then we obtain (5.20) again.   □

Remark 5.17. (a) Composition operators between general Sobolev spaces have been studied by Ukhlov [120]. Our statement of Theorem 5.13 and the detailed proof is by Kleprlík [74].

(b) It was shown in [120] that if f is a homeomorphism and T_f is bounded between the corresponding Sobolev spaces, then $K_q \in L_{loc}^{\frac{p}{q-p}}$. It follows that this condition is not only sufficient but also necessary.

(c) Let us note that classes of mappings for which some ratio of $|Df|$ and J_f is integrable was also studied in papers of Ball. This gives us an independent motivation for the study of the class of mappings with integrable K_q-distortion.

5.3   Sharp Moduli of Continuity for f and f^{-1}

From Theorem 2.4 we know that mappings of finite distortion with locally integrable $\exp(\lambda K)$ are continuous. We establish the following sharp modulus of continuity.

Theorem 5.18. *Let $\Omega \subset \mathbf{R}^n$ be open and let $f : \Omega \to \mathbf{R}^n$ be a homeomorphism of finite distortion. Suppose that there is $\lambda > 0$ such that $\exp(\lambda K_f) \in L_{loc}^1(\Omega)$. For given $\overline{B(z, R)} \subset \Omega$ and $y \in B(z, R)$ we have*

$$|f(y) - f(z)| \leq C_1 \log^{-\lambda/n}\left(\frac{C_2}{|y - z|}\right),$$

where C_1 is the product of $C(n) \operatorname{diam}(f(\overline{B(z, R)}))$ and $\log^{\lambda/n}(\frac{C(n,\lambda)I}{R^n})$, where $I = \int_{B(z,R)} \exp(\lambda K_f)$ and C_2 is $C(n, \lambda)I^{1/n}$.

Proof. Let $\overline{B(z, R)} \subset \Omega$. For simplicity, we assume that $z = 0$. Given $y \in B(0, R)$, set $r = |y|$. Suppose that u is a Lipschitz function so that

$$u = \begin{cases} 1 & \text{on } B(0, r), \\ 0 & \text{on } \Omega \setminus B(0, R) . \end{cases} \tag{5.21}$$

By Theorem 5.9, $f^{-1} \in W^{1,n}(f(B(0, R)), \mathbf{R}^n)$, and hence $u \circ f^{-1} \in W^{1,n}(f(B(0, R)))$. From Theorem A.41 we conclude that

$$\int_{f(B(0,R))} |D(u \circ f^{-1})(a)|^n \, da \geq \omega_{n-1} \log^{1-n}\left(\frac{C(n) \operatorname{diam}(f(B(0, R)))}{\operatorname{diam}(f(B(0, r)))}\right). \tag{5.22}$$

Notice that $|f(y) - f(0)| \leq \operatorname{diam}(f(B(0, r)))$.

Similarly to the proof of Theorem 5.9 we estimate

$$\int_{f(B(0,R))} |D(u \circ f^{-1})(a)|^n \, da \leq \int_{f(B(0,R))} |Du(f^{-1}(a))|^n |Df^{-1}(a)|^n \, da$$

$$= \int_{B(0,R)} |Du(x)|^n |Df^{-1}(f(x))|^n \, J_f(x) \, dx$$

$$= \int_{B(0,R)} |Du(x)|^n \frac{|\operatorname{adj} Df(x)|^n}{J_f(x)^{n-1}} \, dx$$

$$\leq \int_{B(0,R)} |Du(x)|^n \, K_f^{n-1}(x) \, dx, \tag{5.23}$$

and we are reduced to bounding this integral suitably from above in terms of r. To this end, we write $w = K_f^{n-1}$ and define

$$v(t) = \int_r^t \frac{ds}{(\int_{S^{n-1}(0,s)} w \, d\sigma)^{1/(n-1)}}$$

for $r \leq t \leq R$ and further define $u : B(0, R) \setminus \overline{B(0, r)} \to \mathbf{R}$ by setting

$$u(x) = 1 - v(|x|)/v(R).$$

We extend u in the obvious way to the exterior of $B(0, R) \setminus \overline{B(0, r)}$ and we obtain a Lipschitz function (since $K_f \geq 1$, also $w \geq 1$ and it is easy to check that v is Lipschitz) as required in (5.21). We conclude by the Fubini theorem that

$$\int_{B(0,R)} |Du(x)|^n w(x)\, dx \leq \int_r^R \int_{S^{n-1}(0,s)} \left(\frac{1}{v(R)(\int_{S^{n-1}(0,s)} w\, d\sigma)^{1/(n-1)}} \right)^n w\, d\sigma\, ds$$

$$\leq v(R)^{-n} \int_r^R \frac{ds}{(\int_{S^{n-1}(0,s)} w\, d\sigma)^{1/(n-1)}} = v(R)^{1-n}.$$

(5.24)

Hence it suffices to bound $v(R)$ from below.

Next, pick integers i_R and i_r so that $\log R - 1 < i_R \leq \log R$ and $\log r \leq i_r < \log r + 1$. Then

$$v(R) \geq \sum_{i=i_r}^{i_R - 1} \int_{e^i}^{e^{i+1}} \frac{ds}{(\int_{S^{n-1}(0,s)} w\, d\sigma)^{1/(n-1)}}.$$

Now, a change of variable, convexity of $t \to 1/t$ and Jensen's inequality show that

$$\int_{e^i}^{e^{i+1}} \frac{ds}{(\int_{S^{n-1}(0,s)} w\, d\sigma)^{1/(n-1)}} = \int_{e^i}^{e^{i+1}} \frac{ds}{s(\omega_{n-1} \fint_{S^{n-1}(0,s)} w\, d\sigma)^{1/(n-1)}}$$

$$= \int_i^{i+1} \frac{dt}{(\omega_{n-1} \fint_{S^{n-1}(0,e^t)} w\, d\sigma)^{1/(n-1)}}$$

$$\geq \left(\int_i^{i+1} \left(\omega_{n-1} \fint_{S^{n-1}(0,e^t)} w\, d\sigma \right)^{\frac{1}{n-1}} dt \right)^{-1},$$

for each $i_r \leq i \leq i_R - i$. Applying Jensen's inequality again, for the convex functions $t \to \exp(t)$ and $t \to \max\{\exp(n-2), \exp(t^{1/(n-1)})\}$ we see that

$$\int_i^{i+1} \left(\omega_{n-1} \fint_{S^{n-1}(0,e^t)} w\, d\sigma \right)^{\frac{1}{n-1}} dt$$

$$\leq \frac{\omega_{n-1}^{\frac{1}{n-1}}}{\lambda} \log \left(\int_i^{i+1} \exp\left(\left(\fint_{S^{n-1}(0,e^t)} \lambda^{n-1} w\, d\sigma \right)^{\frac{1}{n-1}} \right) dt \right)$$

$$\leq \frac{\omega_{n-1}^{\frac{1}{n-1}}}{\lambda} \log \left(\int_i^{i+1} \fint_{S^{n-1}(0,e^t)} \exp(\lambda \hat{w}^{\frac{1}{n-1}})\, d\sigma\, dt \right),$$

where $\hat{w} = \max\{(n-2)^{n-1}, w\}$. An easy computation shows that

$$\int_{i}^{i+1} \int_{S^{n-1}(0,e^t)} \exp(\lambda \hat{w}^{\frac{1}{n-1}}) \, d\sigma \, dt = \int_{e^i}^{e^{i+1}} \frac{1}{s} \int_{S^{n-1}(0,s)} \exp(\lambda \hat{w}^{\frac{1}{n-1}}) \, d\sigma \, ds$$

$$\leq \frac{1}{\omega_{n-1} e^{in}} \int_{e^i}^{e^{i+1}} \int_{S^{n-1}(0,s)} \exp(\lambda \hat{w}^{\frac{1}{n-1}}) \, d\sigma \, ds$$

$$\leq \frac{CI}{\omega_{n-1} e^{ni}},$$

where $C = 1 + \exp((n-2)\lambda)$. Replacing C by nC we may assume that $\frac{CI}{\omega_{n-1} e^{ni}} \geq 2$ for all $i \leq i_R - 1$. Combining the inequalities above, we conclude that

$$v(R) \geq \sum_{i=i_r}^{i_R-1} \int_{e^i}^{e^{i+1}} \frac{ds}{(\int_{S^{n-1}(0,s)} w(y) \, d\sigma)^{1/(n-1)}}$$

$$\geq \frac{\lambda}{\omega_{n-1}^{\frac{1}{n-1}}} \sum_{i=i_r}^{i_R-1} \log^{-1}\left(\frac{CI}{\omega_{n-1} e^{ni}}\right)$$

$$\geq \frac{\lambda}{\omega_{n-1}^{\frac{1}{n-1}}} \int_r^{R/e^3} \frac{dt}{t \log(\frac{CI}{\omega_{n-1} t^n})}$$

$$\geq \frac{\lambda}{n \omega_{n-1}^{\frac{1}{n-1}}} \log \frac{\log(CI^{1/n}/R)}{\log(C_2 I^{1/n}/r)}.$$

(5.25)

The claim follows by putting the estimates (5.22)–(5.25) together and exponentiating. □

Examples for the Sharp Modulus of Continuity. Extremal mappings for continuity of mappings of finite distortion are usually radial maps and therefore the natural candidate for the extremal map is

$$f_0(x) = \frac{x}{|x|} \frac{1}{\log^{\lambda/n}(1/|x|)}.$$

Standard computations using Theorem 2.1 give us

$$K_f(x) = \frac{n}{\lambda} \log \frac{1}{|x|}$$

and hence

$$\int_{B(0,\frac{1}{2})} \exp(\lambda K_f(x)) \, dx = \int_{B(0,\frac{1}{2})} \frac{1}{|x|^n} \, dx = \infty.$$

In order to obtain the sharp counterexample we need to slightly modify this example.

Theorem 5.19. *Given $\lambda > 0$, there is a mapping of finite distortion $f : B(0, \frac{1}{2}) \to$* $\mathbf{R}^n$ *such that*

$$\int_{B(0,\frac{1}{2})} \exp(\lambda K_f(x))\, dx < \infty$$

and

$$|f(x) - f(0)| \geq \frac{C}{\log^{\lambda/n}(1/|x|)} \text{ for all } x \in B(0, \tfrac{1}{2}). \tag{5.26}$$

Proof. We set

$$f(x) = \frac{x}{|x|} \frac{(\log 1/|x|)^{\frac{a}{\log 1/|x|}}}{\log^{\lambda/n}(1/|x|)}$$

where $a > 0$. The additional term clearly satisfies

$$\lim_{|x|\to 0} (\log 1/|x|)^{\frac{a}{\log 1/|x|}} = 1$$

and thus the modulus of continuity of our f is exactly as required in (5.26). On the other hand the additional term slightly affects the distortion.

Using Lemma 2.1 we get

$$|Df(x)| = \frac{(\log 1/|x|)^{\frac{a}{\log 1/|x|}}}{|x| \log^{\lambda/n}(1/|x|)} \max\left\{1, \left|\frac{\frac{\lambda}{n}}{\log(1/|x|)} + \left(\frac{a \log\log \frac{1}{|x|}}{\log^2 \frac{1}{|x|}} - \frac{a}{\log^2 \frac{1}{|x|}}\right)\right|\right\}.$$

Since

$$\lim_{x\to 0}\left[\frac{\frac{\lambda}{n}}{\log(1/|x|)} + \frac{a \log\log \frac{1}{|x|}}{\log^2 \frac{1}{|x|}} - \frac{a}{\log^2 \frac{1}{|x|}}\right] = 0$$

we obtain for small enough $|x|$ using Lemma 2.1 and elementary inequalities that

$$K_f(x) = \frac{1}{\frac{\frac{\lambda}{n}}{\log(1/|x|)} + \frac{a(\log\log \frac{1}{|x|}-1)}{\log^2 \frac{1}{|x|}}} \leq \frac{n}{\lambda} \log \frac{1}{|x|} - \frac{n^2 a}{2\lambda^2} \log\log \frac{1}{|x|}.$$

Since the $\exp(\lambda \cdot)$ of the last term corresponds to $\dfrac{1}{|x|^n \log^{a\frac{n^2}{2\lambda^2}} \frac{1}{|x|}}$ we obtain that for sufficiently large a

$$\int_{B(0,\frac{1}{2})} \exp(\lambda K_f(x))\, dx < \infty. \qquad \square$$

Modulus of Continuity for the Inverse. Recall from Theorem 5.9 that $K_f \in L_{loc}^{n-1}(\Omega)$ implies that $f^{-1} \in W_{loc}^{1,n}(f(\Omega), \mathbf{R}^n)$. By Remark 2.22, this gives a locally uniform modulus of continuity of logarithmic type for f^{-1}.

Our next result gives an essentially sharp modulus of continuity for f^{-1} when $K_f \in L_{loc}^p(\Omega)$ for some $p > n - 1$.

Theorem 5.20. *Let $\Omega \subset \mathbf{R}^n$ be an open set and let $f : \Omega \to f(\Omega)$ be a homeomorphism of finite distortion with $K_f \in L_{loc}^p(\Omega)$ for some $p > n - 1$. Let $B(z, R) \subset\subset f(\Omega)$. Then for all points $x \in B(z, R)$,*

$$|f^{-1}(x) - f^{-1}(z)| \leq C(p,n) \|K_f\|_{L^p(f^{-1}(B(z,R)))}^{\frac{p}{n}} \log^{-\frac{p(n-1)}{n}} \left(\frac{R}{|x - z|} \right).$$

Moreover, the exponent $-\frac{p(n-1)}{n}$ is optimal.

Proof. Set $q = \frac{pn}{p+1}$; then $n - 1 < q < n$. Let $x \in B(z, R)$, where $B(z, R)$ is in our assumptions, and set E to be the line segment between z and x. By Theorem A.41, we find a Lipschitz function u so that $u = 1$ on E, $u = 0$ on $f(\Omega) \setminus B(z, R)$, and

$$\int_{f(\Omega)} |Du|^n \leq \omega_{n-1} \log^{1-n} \left(\frac{R}{|x - z|} \right). \tag{5.27}$$

Put $v := u \circ f$. Then $v \in W^{1,1}(\Omega)$ and the chain rule in conjunction with the distortion inequality imply that

$$|Dv|^q \leq |Du(f)|^q K_f^{\frac{q}{n}} J_f^{\frac{q}{n}},$$

where $\frac{q}{n} = \frac{p}{p+1}$. We use Hölder's inequality and a change of variables to get

$$\int_{f^{-1}(B(z,R))} |Dv|^q \leq \int_{f^{-1}(B(z,R))} (|Du(f)|^n J_f)^{\frac{q}{n}} K_f^{\frac{q}{n}}$$

$$\leq \left(\int_{f^{-1}(B(z,R))} |Du(f)|^n J_f \right)^{\frac{q}{n}} \left(\int_{f^{-1}(B(z,R))} (K_f^{\frac{q}{n}})^{\frac{n}{n-q}} \right)^{\frac{n-q}{n}}$$

$$\leq \left(\int_{B(z,R)} |Du|^n \right)^{\frac{p}{p+1}} \left(\int_\Omega K_f^p \right)^{\frac{1}{p+1}}$$

$$= \|K_f\|_{L^p(f^{-1}(B(z,R)))}^{\frac{p}{p+1}} \left(\int_{B(z,R)} |Du|^p \right)^{\frac{p}{p+1}}. \tag{5.28}$$

On the other hand, Corollary A.40 gives us that

$$(\operatorname{diam}(f^{-1}(E)))^{n-q} \leq C(p,n) \int_{f^{-1}(B(z,R))} |Dv|^q. \tag{5.29}$$

Our first claim follows by combining (5.27)–(5.29), once we notice that
$\text{diam}(f^{-1}(E)) \geq |f^{-1}(x) - f^{-1}(z)|$.

Regarding the sharpness of the exponent $-\frac{p(n-1)}{n}$, we show that the estimate
fails to hold for any smaller exponent. To this end, let $\alpha > \frac{p(n-1)}{n}$. Choose $\beta \in$
$(\frac{1}{\alpha}, \frac{n}{p(n-1)})$. We define a radial homeomorphism $f : \mathbf{R}^n \to \mathbf{R}^n$ by setting $f(0) = 0$
and

$$f(x) := \rho(|x|)\frac{x}{|x|}$$

for $x \neq 0$, where $\rho(t) := \exp(-t^{-\beta})$. By Lemma 2.1 we see that

$$K_f(x) = \left(\frac{\beta}{|x|^\beta}\right)^{n-1}$$

when $|x| \leq \beta$. Since $\beta(n-1)p < n$, we conclude that $K_f \in L^p(B(0, \beta))$. On the
other hand,

$$|f^{-1}(0) - f^{-1}(y)| = \log^{-\frac{1}{\beta}}(1/|y|),$$

and hence the desired estimate does not hold with the exponent $-\alpha$. $\square$

Remark 5.21. (a) The sharp modulus of continuity in Theorem 5.18 was obtained
   by Onninen and Zhong in [101]; the proof above is based on the approach of
   Koskela and Onninen [78].
(b) The sharp modulus of continuity in Theorem 5.20 was established by Clop and
   Herron in [19] and in dimension two this was already given in [81].
(c) The example for the sharp modulus of continuity in Theorem 5.19 is from
   Campbell and Hencl [14].
(d) Under the assumption that $\exp(\lambda K_f) \in L^1$ the modulus of continuity in
   Theorem 5.20 improves to a function of the form $\exp(-C \log^{\frac{n}{n-1}} 1/t)$, see [59].

Open problem 13. Notice that the exponent $-\frac{p(n-1)}{n}$ from Theorem 5.20 tends to
$-\frac{(n-1)^2}{n}$ when p tends to $n - 1$. In the borderline case, we know from Theorem 5.9
that $f^{-1} \in W^{1,n}_{\text{loc}}$ and it follows by Remark 2.22 that we have a logarithmic modulus
of continuity with the exponent $-\frac{1}{n}$. The former exponent is the optimal one for
radial mappings as in the proof of Theorem 5.20. These two exponents do not
coincide and we would like to know the optimal exponent in this borderline case.

5.4 Jacobian Cannot Change Sign

In this section, we address the following problem, originally asked by P. Hajlasz.
Suppose that $\Omega \subset \mathbf{R}^n$ is a domain and that $f : \Omega \to \mathbf{R}^n$ is a homeomorphism of
the Sobolev class $W^{1,1}_{\text{loc}}(\Omega, \mathbf{R}^n)$. Is it true that the Jacobian J_f is either non-negative

almost everywhere or non-positive almost everywhere? It is well-known that every homeomorphism defined on a domain Ω is either sense-preserving or sense-reversing and therefore we can ask whether each sense-preserving homeomorphism in the Sobolev space $W_{loc}^{1,s}$ satisfies $J_f \geq 0$ almost everywhere. Roughly speaking, we are interested in the question whether topological and analytical definitions of orientation lead to the same result.

The following theorem tells us that the answer to our problem is in the positive if our mapping is differentiable in the classical sense. Recall that by Corollary 2.25 we know that each homeomorphism $f \in W^{1,p}(\Omega, \mathbf{R}^n)$, $p > n - 1$ for $n > 2$ or $p \geq 1$ for $n = 2$ is differentiable a.e. Thus the following result gives us a positive answer to our problem for $n = 2$.

Theorem 5.22. *Let $\Omega \subset \mathbf{R}^n$ be domain and suppose that $f : \Omega \to f(\Omega)$ is a homeomorphism with $f \in W_{loc}^{1,1}(\Omega, \mathbf{R}^n)$ and that it is differentiable a.e. Then either $J_f \geq 0$ a.e. or $J_f \leq 0$ a.e.*

Proof. Let us fix a domain $\Omega' \subset\subset \Omega$. We will show that the sign of J_f is constant a.e. on Ω'. Since $f(\Omega')$ is connected, the topological degree with respect to $\mathbf{R}^n \setminus f(\partial\Omega')$ is constant on $f(\Omega')$.

Let us consider a point $x_0 \in \Omega'$ such that f is differentiable at x_0 and $J_f(x_0) \neq 0$. Then $\alpha = \inf_{\|y\|=1} |Df(x_0)y| > 0$ and we can choose $r > 0$ small enough such that

$$|f(x_0 + x) - f(x_0) - Df(x_0)x| < r\alpha \text{ for every } x \in S^{n-1}(0, r). \qquad (5.30)$$

Let us consider the homotopy

$$H(x, t) = (1 - t)\big(f(x_0 + x) - f(x_0)\big) + tDf(x_0)x \text{ for } x \in B(0, r).$$

By (5.30) we obtain that $0 \notin H(S^{n-1}(0, r), t)$ for every $t \in [0, 1]$. From Remark 3.12 (f) we know that the degree is stable under homotopy and thus

$$\deg(f(x_0), f(x), B(x_0, r)) = \deg(0, f(x_0 + x) - f(x_0), B(0, r))$$
$$= \deg(0, Df(x_0)x, B(0, r)) = \text{sgn } J_f(x_0).$$

The degree is constant on $f(\Omega')$ and hence the sign of the Jacobian cannot change. □

Theorem 5.22 was improved by Hencl and Malý in [53] where they used the stability of the linking number and proved the following. Note that this gives the full answer to our problem in dimension $n = 3$.

Theorem 5.23. *Let $\Omega \subset \mathbf{R}^n$, $n \geq 2$, be a domain, $p = 1$ for $n = 2, 3$ and $p > [n/2]$ for $n \geq 4$. Suppose that $f \in W_{loc}^{1,p}(\Omega, \mathbf{R}^n)$ is homeomorphism. Then either $J_f \geq 0$ a.e. or $J_f \leq 0$ a.e.*

Open problem 14. Let $\Omega \subset \mathbf{R}^n$, $n \geq 4$, be a domain. Suppose that $f \in W^{1,1}(\Omega, \mathbf{R}^n)$ is a homeomorphism. Is it true that either $J_f \geq 0$ a.e. or $J_f \leq 0$ a.e.? The same question can be asked also for $f \in BV(\Omega, \mathbf{R}^n)$, $n \geq 4$. In dimensions $n = 2, 3$ the previous proofs can be carried out with small modifications also for BV.

5.5   Approximation of Sobolev Homeomorphisms

Let us close this section by one of the most interesting and important problems in this area, originally posed by Evans and later promoted by Ball [11], [10].

Open problem 15 ([10]). Let $\Omega \subset \mathbf{R}^n$ be a domain and $1 \leq p < \infty$. Suppose that $f \in W^{1,p}(\Omega, \mathbf{R}^n)$ is a homeomorphism. Is it possible to find a sequence of piecewise affine homeomorphisms $\tilde{f}_k$ such that $\|\tilde{f}_k - f\|_{W^{1,p}} \to 0$? Is it possible to find a sequence of smooth homeomorphisms f_k such that $\|f_k - f\|_{W^{1,p}} \to 0$?

Partial motivation for this problem comes from regularity of models in nonlinear elasticity. Also, piecewise affine approximation would be nice for numerical approximation. We recommend [10, 11] for more on this subject.

This problem is nontrivial because the usual approximation techniques like mollification can well destroy the injectivity of a given homeomorphism. It is easy to see that if we can construct smooth f_k that approximate f, then we can also construct piecewise linear $\tilde{f}_k$. On the other hand, in dimensions two and three, the existence of piecewise affine approximations implies the existence of smooth approximations ([66],[97]).

The first positive approximation result for mappings that are smooth outside a single point was given by Mora-Corral [96]. The following deep result of Iwaniec et al. [65, 66] essentially solves the problem in dimension $n = 2$.

Theorem 5.24. *Let $\Omega \subset \mathbf{R}^2$ be domain and let $1 < p < \infty$. Suppose that $f \in W^{1,p}(\Omega, \mathbf{R}^2)$ is a homeomorphism. Then there are smooth homeomorphisms f_k such that $f_k - f \in W_0^{1,p}(\Omega, \mathbf{R}^2)$, $f_k \to f$ uniformly and $\|Df - Df_k\|_{L^p} \to 0$.*

Unfortunately the method of the proof heavily relies on the properties of the p-Laplace equation in the plane. Thus the higher dimensional setting appears to require a new technique. For some results in the case $p = 1$ and $n = 2$ see Hencl and Pratelli [56].

In models of nonlinear elasticity one usually knows that also the inverse mapping is weakly differentiable and thus the following problem is natural.

Open problem 16 ([65]). Let $\Omega \subset \mathbf{R}^n$ be a domain and $1 \leq p < \infty$. Suppose that $f \in W^{1,p}(\Omega, \mathbf{R}^n)$ is a p-bi-Sobolev homeomorphism, i.e. $f^{-1} \in W^{1,p}(f(\Omega), \mathbf{R}^n)$. Is it possible to find a sequence of piecewise affine homeomorphisms $\tilde{f}_k$ such that $\|\tilde{f}_k - f\|_{W^{1,p}} \to 0$ and $\|\tilde{f}_k^{-1} - f^{-1}\|_{W^{1,p}} \to 0$? Is it possible to find a sequence of smooth homeomorphisms f_k such that $\|f_k - f\|_{W^{1,p}} \to 0$ and $\|f_k^{-1} - f^{-1}\|_{W^{1,p}} \to 0$?

This problem was solved in the plane $n = 2$ under the additional assumption that f is bi-Lipschitz by Daneri and Pratelli [21]. The solution was based on their earlier result [22] according to which each bi-Lipschitz mapping, defined on the boundary of a cube, can be extended to a bi-Lipschitz mapping defined on the entire cube. For and overview of the planar results and methods we recommend the expository article of Pratelli and Puglisi [106].

Chapter 6
Integrability of J_f and $1/J_f$

Abstract In this chapter we study the optimal degree of integrability of J_f and $1/J_f$ for mappings of finite distortion. As an application of our estimates we show that some sets are removable singularities for mappings with exponentially integrable distortion.

6.1 Regularity of the Jacobian under $\exp(\lambda K) \in L^1$

In Sect. 1.2, we recalled that for each quasiregular mapping $f \in W^{1,n}(\Omega, \mathbf{R}^n)$ there is $p > n$ such that $|Df| \in L^p_{\text{loc}}(\Omega)$. This remarkable self-improving regularity result, which is based on a reverse Hölder inequality, is important for many other properties of quasiregular mappings.

In Lemma 2.8 we saw that each mapping of finite distortion with exponentially integrable distortion naturally satisfies $|Df| \in L^n \log^{-1} L$. Actually, the situation is slightly better.

Theorem 6.1. *Let $\Omega \subset \mathbf{R}^n$, $n \geq 2$, be a domain and let $f \in W^{1,1}_{\text{loc}}(\Omega, \mathbf{R}^n)$ be a mapping of finite distortion. Assume that $\exp(\beta K_f) \in L^1_{\text{loc}}(\Omega)$, for some $\beta > 0$. Then*

$$J_f \log^\alpha(e + J_f) \in L^1_{\text{loc}}(\Omega), \quad \text{and} \quad |Df|^n \log^{\alpha-1}(e + |Df|) \in L^1_{\text{loc}}(\Omega),$$

where $\alpha = C_1\beta$ and $C_1 = C_1(n) > 0$. Moreover, for any ball B such that $2B \subset\subset \Omega$,

$$\fint_B J_f(x) \log^\alpha\left(e + \frac{J_f(x)}{(J_f)_{2B}}\right) dx \leq C(n, \beta)\left(\fint_{2B} \exp(\beta K(x)) \, dx\right)\left(\fint_{2B} J_f(x) \, dx\right).$$

$$(6.1)$$

From Theorem A.6 and Lemma A.12 we know that the maximal function of $h \in L^1(\mathbf{R}^n)$ satisfies

S. Hencl and P. Koskela, *Lectures on Mappings of Finite Distortion*, Lecture Notes in Mathematics 2096, DOI 10.1007/978-3-319-03173-6_6,
© Springer International Publishing Switzerland 2014

$$\frac{1}{2^n t} \int_{\{|h|>t\}} |h(x)| \, dx \le |\{x \in \mathbf{R}^n : Mh(x) > t\}| \le \frac{2 \cdot 3^n}{t} \int_{\{|h|>t/2\}} |h(x)| \, dx$$

$$(6.2)$$

for every $t > 0$ which will be essential for our proof.

Lemma 6.2. *Let $\alpha > 0$ and $\beta > 0$. For every $a \ge 1$ and $b \ge 1$ we have*

$$ab \log^{\alpha-1}(C(n)(ab)^{\frac{1}{n}}) \le \frac{C(n)}{\beta} a \log^{\alpha}(a^{\frac{1}{n}}) + C(\alpha, \beta, n) \exp(\beta b).$$

Proof. The case $a \le e$ is easy. Suppose $a > e$. We have that

$$ab \log^{-1}(C(n)(ab)^{\frac{1}{n}}) \le ab \log^{-1}(a^{\frac{1}{n}})$$

$$\le \left(\frac{4}{\beta} a \log a + \exp(\frac{\beta}{2} b)\right) \log^{-1} a^{\frac{1}{n}} \qquad (6.3)$$

$$\le C(n) \left(\frac{1}{\beta} a + \exp(\frac{\beta}{2} b)\right),$$

where we used the elementary inequality

$$ab \le a \log a + \exp(2b)$$

for $a \ge 1, b \ge 1$. We also have that

$$\log^{\alpha}(C(n)(ab)^{\frac{1}{n}}) \le 2 \log^{\alpha} a^{\frac{1}{n}} + C(n, \alpha) \log^{\alpha}(C(n)b), \qquad (6.4)$$

which follows from

$$(x + y)^{\alpha} \le 2x^{\alpha} + C(\alpha)y^{\alpha}$$

for $x > 0, y > 0, \alpha > 0$.

Combining (6.3) and (6.4) yields

$$ab \log^{\alpha-1}(C(n)(ab)^{\frac{1}{n}}) \le C(n)\left(\frac{1}{\beta} a + \exp(\frac{\beta}{2} b)\right)\left(2 \log^{\alpha} a^{\frac{1}{n}} + C(n, \alpha) \log^{\alpha}(C(n)b)\right)$$

$$\le C(n) \left(\frac{1}{\beta} a \log^{\alpha} a^{\frac{1}{n}} + C(n, \alpha, \beta) \exp(\beta b)\right), \qquad (6.5)$$

as desired. The last inequality holds because of the estimates

$$C(n, \alpha)a \log^{\alpha}(C(n)b) \le a \log^{\alpha} a^{\frac{1}{n}} + C(n, \alpha, \beta) \exp(\beta b),$$

and

$$\exp\left(\frac{\beta}{2}b\right) \log^\alpha a^{\frac{1}{n}} \le a \log^\alpha a^{\frac{1}{n}} + C(n, \alpha, \beta) \exp(\beta b),$$

for the lower order terms which can be easily proven.   □

Proof (of Theorem 6.1). From Theorem 2.12 we know that the distributional Jacobian coincides with the pointwise Jacobian, i.e.

$$\int_\Omega \varphi(x) J_f(x) \, dx = -\int_\Omega f_1(x) J(\varphi, f_2, \ldots, f_n)(x) \, dx$$

whenever $\varphi \in C_0^\infty(\Omega)$. Thus

$$\int_\Omega \varphi(x) J_f(x) \, dx \le \int_\Omega |f(x)| |Df(x)|^{n-1} |\nabla\varphi(x)| \, dx. \tag{6.6}$$

This is a reverse inequality, from which the higher integrability result will be derived. Let $\varphi \in C_0^\infty(B(y, 2r))$ satisfy $\varphi = 1$ in $B(y, r)$, $0 \le \varphi \le 1$ in $\mathbf{R}^n$ and $|\nabla\varphi| \le 2/r$, where $B(y, 2r)$ is a ball in Ω. With this choice of φ, (6.6) leads to the following inequality with $q = n^2/(n+1)$:

$$\int_{B(y,r)} J_f(x) \, dx \le \frac{2}{r} \int_{B(y,2r)} |f(x)| |Df(x)|^{n-1} \, dx$$

$$\le \frac{2}{r} \left(\int_{B(y,2r)} |Df(x)|^q \, dx \right)^{\frac{n-1}{q}} \left(\int_{B(y,2r)} |f(x)|^{n^2} \, dx \right)^{\frac{1}{n^2}}.$$

Note that this inequality remains valid if we subtract from f any constant vector. In particular, it holds for $f - f_{B(y,2r)}$ replacing f. The Poincaré-Sobolev inequality Theorem A.18 yields

$$\left(\int_{B(y,2r)} |f(x) - f_{B(y,2r)}|^{n^2} \, dx \right)^{\frac{1}{n^2}} \le C(n) \left(\int_{B(y,2r)} |Df(x)|^q \, dx \right)^{\frac{1}{q}}.$$

Combining these last two inequalities, we finally obtain

$$\frac{1}{|B(y,r)|} \int_{B(y,r)} J_f(x) \, dx \le C(n) \left(\frac{1}{|B(y,2r)|} \int_{B(y,2r)} |Df(x)|^q \, dx \right)^{\frac{n}{q}} \tag{6.7}$$

whenever $B(y, 2r) \subset\subset \Omega$.

Now we fix a ball $B_0 = B(x_0, r_0) \subset\subset \Omega$. Assume that

$$\int_{B_0} J_f(x) \, dx = 1. \tag{6.8}$$

This assumption involves no loss of generality for us as the distortion inequality and (6.1) are homogeneous with respect to f. Let us introduce the auxiliary functions defined in $\mathbf{R}^n$ by (abuse of notation)

$$h_1(x) = d(x)^n J_f(x),$$
$$h_2(x) = d(x)|Df(x)|, \qquad (6.9)$$
$$h_3(x) = \chi_{B_0}(x),$$

where $d(x) = \mathrm{dist}(x, \mathbf{R}^n \setminus B_0)$. We claim that

$$\left(\frac{1}{|B|}\int_B h_1\,dx\right)^{\frac{1}{n}} \le C(n)\left(\frac{1}{|2B|}\int_{2B} h_2^q\,dx\right)^{\frac{1}{q}} + C(n)\left(\frac{1}{|2B|}\int_{2B} h_3\,dx\right)^{\frac{1}{n}} \quad (6.10)$$

for all balls $B \subset \mathbf{R}^n$. Indeed, we may assume that B meets B_0; otherwise (6.10) is trivial. Our derivation of (6.10) naturally falls into two cases.

Case 1.   We assume that $3B \subset B_0$. By an elementary geometric consideration we find that

$$\max_{x \in B} d(x) \le 4 \min_{x \in 2B} d(x).$$

Applying (6.7) yields

$$\left(\frac{1}{|B|}\int_B h_1\,dx\right)^{\frac{1}{n}} \le \max_B d(x)\left(\frac{1}{|B|}\int_B J_f(x)\,dx\right)^{\frac{1}{n}}$$

$$\le C(n)\min_{2B} d(x)\left(\frac{1}{|2B|}\int_{2B}|Df(x)|^q\,dx\right)^{\frac{1}{q}}$$

$$\le C(n)\left(\frac{1}{|2B|}\int_{2B} h_2^q\,dx\right)^{\frac{1}{q}}.$$

Case 2.   We assume that $3B$ is not contained in B_0 and recall that B meets B_0. We have that

$$\max_{x \in B} d(x) \le \max_{x \in 2B} d(x) \le C(n)|2B \cap B_0|^{\frac{1}{n}}.$$

Hence we conclude that

$$\left(\frac{1}{|B|}\int_B h_1\,dx\right)^{\frac{1}{n}} \le \max_B d(x)\left(\frac{1}{|B|}\int_{B\cap B_0} J_f(x)\,dx\right)^{\frac{1}{n}}$$

$$\le C(n)\left(\frac{|2B \cap B_0|}{|B|}\int_{B_0} J_f(x)\,dx\right)^{\frac{1}{n}}$$

$$\le C(n)\left(\frac{1}{|2B|}\int_{2B} h_3\,dx\right)^{\frac{1}{n}},$$

where we used (6.8). Combining these two cases proves inequality (6.10).

Since (6.10) is true for all balls $B \subset \mathbf{R}^n$, we have the following pointwise inequality for the maximal functions. For all $y \in \mathbf{R}^n$,

$$M(h_1)(y)^{\frac{1}{n}} \leq C(n)M(h_2^q)(y)^{\frac{1}{q}} + C(n)M(h_3)(y)^{\frac{1}{n}},$$

from which it follows that for $\lambda > 0$

$$|\{x \in \mathbf{R}^n : M(h_1)(x) > \lambda^n\}| \leq |\{x \in \mathbf{R}^n : C(n)M(h_2^q)(x) > \lambda^q\}|$$
$$+ |\{x \in \mathbf{R}^n : C(n)M(h_3)(x) > \lambda^n\}|.$$

We recall that $h_3(x) = \chi_{B_0}(x)$. So $M(h_3)(x) \leq 1$ in $\mathbf{R}^n$, and then the set $\{x \in \mathbf{R}^n : c(n)M(h_3)(x) > \lambda^n\}$ is empty for $\lambda > \lambda_1 = \lambda_1(n)$. Hence

$$|\{x \in \mathbf{R}^n : M(h_1)(x) > \lambda^n\}| \leq |\{x \in \mathbf{R}^n : C(n)M(h_2^q)(x) > \lambda^q\}|$$

for all $\lambda > \lambda_1$. Now applying (6.2) yields

$$\int_{\{h_1 > \lambda^n\}} h_1 \, dx \leq C(n)\lambda^{n-q} \int_{\{C(n)h_2 > \lambda\}} h_2^q \, dx \tag{6.11}$$

for all $\lambda > \lambda_1$. We may assume that the constant $C(n)$ in (6.11) is bigger than one.

Let $\alpha > 0$ be a constant, which will be chosen later, and set

$$\Psi(\lambda) = \frac{n-q}{\alpha} \log^\alpha \lambda + \log^{\alpha-1} \lambda,$$

where $q = n^2/(n+1)$ as above. Notice that

$$\Phi(\lambda) := \frac{d}{d\lambda} \Psi(\lambda) = \frac{n-q}{\lambda} \log^{\alpha-1} \lambda + \frac{\alpha-1}{\lambda} \log^{\alpha-2} \lambda > 0$$

for all $\lambda > \lambda_2 = \exp((n+1)/n)$, and that

$$\lambda^{n-q} \Phi(\lambda) = \frac{d}{d\lambda} \left(\lambda^{n-q} \log^{\alpha-1} \lambda \right).$$

We multiply both sides of (6.11) by $\Phi(\lambda)$, and integrate with respect to λ over (λ_0, ∞) for $\lambda_0 = \max\{\lambda_1, \lambda_2\}$, and finally change the order of the integration to obtain that

$$\int_{\{h_1 > \lambda_0^n\}} h_1 \int_{\lambda_0}^{h_1^{\frac{1}{n}}} \Phi(\lambda) \, d\lambda dx \leq C(n) \int_{\{C(n)h_2 > \lambda_0\}} h_2^q \int_{\lambda_0}^{C(n)h_2} \lambda^{n-q} \Phi(\lambda) \, d\lambda dx,$$

that is,

$$\int_{\{h_1 > \lambda_0^n\}} \left(\Psi(h_1^{\frac{1}{n}}) - \Psi(\lambda_0) \right) h_1 \, dx \le C(n) \int_{\{C(n)h_2 > \lambda_0\}} h_2^n \log^{\alpha-1}(C(n)h_2) \, dx .$$

Hence, taking into account the normalization (6.8),

$$\frac{1}{\alpha} \int_{\{h_1 > \lambda_0^n\}} h_1 \log^\alpha h_1^{\frac{1}{n}} \, dx \le C(n) \int_{\{C(n)h_2 > \lambda_0\}} h_2^n \log^{\alpha-1}(C(n)h_2) \, dx + C(n,\alpha)|B_0|,$$

$$(6.12)$$

where $C(n) \ge 1$.

In the remaining part of the proof, we will choose a suitable constant $\alpha > 0$ so that the integral on the right-hand side of (6.12) can be absorbed in the left, by using the distortion inequality. Actually, this only works if we have a priori $|Df| \in L^n \log^{\alpha-1} L_{\text{loc}}(\Omega)$. We cannot assume this. To overcome this, in the above argument we integrate with respect to λ over (λ_0, j) for j large, instead of over (λ_0, ∞). Then we proceed in the same way as above and obtain an inequality similar to (6.12). The proof will be then eventually concluded, by letting $j \to \infty$ and using the monotone convergence theorem. For simplicity, we only write down the proof for $j = \infty$.

We recall the definitions of h_1 and h_2 in (6.9) and notice that the distortion inequality gives

$$h_1(x) \le h_2^n(x) \le h_1(x) K_f(x). \qquad (6.13)$$

We use (6.12), (6.13) and Lemma 6.2 with $a = h_1(x)$ and $b = K_f(x)$ to conclude that

$$\frac{1}{\alpha} \int_{\{h_1 > \lambda_0^n\}} h_1 \log^\alpha h_1^{\frac{1}{n}} \, dx \le \frac{C(n)}{\beta} \int_{\{h_1 > \lambda_0^n\}} h_1 \log^\alpha h_1^{\frac{1}{n}} \, dx$$

$$+ C(n,\alpha,\beta) \int_{B_0} \exp(\beta K(x)) \, dx + C(n,\alpha)|B_0|$$

$$\le \frac{C(n)}{\beta} \int_{\{h_1 > \lambda_0^n\}} h_1 \log^\alpha h_1^{\frac{1}{n}} \, dx$$

$$+ C(n,\alpha,\beta) \int_{B_0} \exp(\beta K(x)) \, dx. \qquad (6.14)$$

Now letting $\alpha = \beta/(2C(n))$, (6.14) becomes

$$\int_{\{h_1 > \lambda_0^n\}} h_1 \log^\alpha h_1^{\frac{1}{n}} \, dx \le C(n,\beta) \int_{B_0} \exp(\beta K(x)) \, dx.$$

That is,

$$\int_{B_0} d(x)^n J_f(x) \log^\alpha (e + d(x)^n J_f(x))\, dx = \int_{B_0} h_1 \log^\alpha (e + h_1)\, dx$$

$$\leq C(n, \beta) \int_{B_0} \exp(\beta K(x))\, dx.$$

Noticing that in $\sigma B_0 = B(x_0, \sigma r_0)$ with $0 < \sigma < 1$ we have $d(x)^n \geq (1 - \sigma)^n r_0^n \geq C(n, \sigma)|B_0|$, and taking account of the normalization (6.8), we arrive at

$$\fint_{\sigma B_0} J_f(x) \log^\alpha \left(e + \frac{J_f(x)}{\fint_{B_0} J_f(x)\, dx}\right) dx$$

$$\leq C(n, \beta, \sigma) \left(\fint_{B_0} \exp(\beta K(x))\, dx\right) \left(\fint_{B_0} J_f(x)\, dx\right),$$

which proves (6.1). The $L^n \log^{\alpha-1} L$-integrability of $|Df|$ follows by the distortion inequality and Lemma 6.2 with $a = K_f^{\frac{1}{n}}$ and $b = J_f^{\frac{1}{n}}$; we use the exponential integrability of K_f and $L^1 \log^\alpha L$-integrability of the Jacobian established above.
$\square$

Remark 6.3. (a) The results of this section were established by Faraco et al. in [30]. For earlier related results see [60, 61].
(b) It was shown by Astala, Gill, Saksman and Rohde in [7] that Theorem 6.1 holds in the plane $n = 2$ with any constant $C_1 > 1$ but it does not hold with $C_1 = 1$.

Open problem 17. Find the optimal value of C_1 in dimensions $n \geq 3$.

6.2 Integrability of $1/J_f$

Recall from Sect. 1.2 that a small negative power of J_f is integrable if f is a non-constant mapping of bounded distortion. Based on Theorem 4.13 we know that already the local integrability of $K_f^{\frac{1}{n-1}}$ is sufficient to guarantee that $J_f(x) > 0$ almost everywhere, provided f is homeomorphic. It is thus natural to ask if $1/J_f$ is integrable to some extent. The following result from [79] gives the optimal conclusion.

Theorem 6.4. *Let $f : \Omega \to f(\Omega) \subset \mathbf{R}^n$ be a homeomorphic mapping of finite distortion so that for some $p \geq 1$ we have*

$$K_f^{\frac{1}{n-1}} \in L_{loc}^p(\Omega).$$

Then

$$\log\left(e + \frac{1}{J_f(\cdot)}\right) \in L^p_{\mathrm{loc}}(\Omega).$$

Example 6.5. Let $\varepsilon > 0$. Set

$$f(x) = \frac{x}{|x|}\exp\left(-\frac{1}{|x|^{\frac{n}{p+\varepsilon}}}\right).$$

By Lemma 2.1 we obtain that for small enough $|x|$ we have

$$K_f(x) = \left(\frac{n}{p+\varepsilon}\frac{1}{|x|^{\frac{n}{p+\varepsilon}}}\right)^{n-1} \text{ and } J_f(x) = \exp\left(-\frac{n}{|x|^{\frac{n}{p+\varepsilon}}}\right)\frac{n}{p+\varepsilon}\frac{1}{|x|^{n-1+\frac{n}{p+\varepsilon}}}.$$

It follows that

$$K_f^{\frac{1}{n-1}} \in L^p, \text{ but } \log\left(e + \frac{1}{J_f}\right) \approx \frac{1}{|x|^{\frac{n}{p+\varepsilon}}} \notin L^{p+\varepsilon}.$$

For an even more optimal example see [79, Example 1.1].   □

Let us briefly discuss the planar setting. Then also $g = f^{-1}$ has finite distortion by Theorem 1.6 and, via a change of variables, the claim reduces to showing that $J_g \log^p(e + J_g) \in L^1_{\mathrm{loc}}(f(\Omega))$. Towards this end, recall the argument from the previous section for a similar conclusion. The crucial point there was to apply a suitable reverse Hölder inequality that was obtained via the coincidence of the pointwise and distributional Jacobians. In the current setting, these two Jacobians need not coincide, but we have

$$\fint_B J_g \leq C\left(\fint_{2B} |Dg|\right)^2, \tag{6.15}$$

whenever B is a ball with $2B \subset f(\Omega)$. Indeed, this follows from the usual isoperimetric inequality via a change of variables. Hence there is hope that a suitable modification to the arguments from the previous section could give our claim. This is indeed the case.

The higher dimensional setting is harder: one does not necessarily have that $f^{-1} \in W^{1,1}_{\mathrm{loc}}$. However, one still has a version of (6.15). For this and for a detailed proof see [79], where the claim of Theorem 6.4 is actually established for open and discrete mappings instead of homeomorphisms. This causes additional difficulties.

Open problem 18. Suppose that f is a non-constant mapping of finite distortion with $\exp(\lambda K_f) \in L^1_{\mathrm{loc}}(\Omega)$ for some $\lambda > 0$. What is the optimal degree of integrability of $1/J_f$? This is open even in the homeomorphic case.

6.3 Application to Removable Singularities

In Theorem 6.1 we have seen that mappings with exponentially integrable distortion (and $J_f \in L^1$) have better regularity than $|Df| \in L^n \log^{-1} L$. In this section we find conditions that guarantee that our mappings satisfy the natural integrability condition $|Df| \in L^n \log^{-1} L$ and $J_f \in L^1_{loc}(\Omega)$ which are crucial for many positive results.

Theorem 6.6. *Let* $f \in W^{1,1}_{loc}(\Omega, \mathbf{R}^n)$, $n \geq 2$, *satisfy the distortion inequality*

$$|Df(x)|^n \leq K(x)J_f(x) \qquad \text{a.e. in } \Omega$$

where $K(x) \geq 1$ *satisfies* $\exp(\beta K) \in L^1_{loc}(\Omega)$, *for some* $\beta > 0$. *There is a constant* $C_2 = C_2(n)$ *such that if*

$$|Df|^n \log^{-\alpha - 1}(e + |Df|) \in L^1_{loc}(\Omega),$$

with $\alpha = C_2\beta$, *then* $|Df| \in L^n \log^{-1} L_{loc}(\Omega)$ *and* $J_f(x) \in L^1_{loc}(\Omega)$. *In particular,* f *is then a mapping of finite distortion.*

Using Lemma 2.1 it is easy to see that the mapping $f(x) = \log^s(e + 1/|x|)\frac{x}{|x|}$, defined in the unit ball of $\mathbf{R}^n$, for which $J_f(x)$ is not locally integrable when $s > 0$, shows that for each $C_2 < 1$ and any β, there are mappings for which the claim fails. Thus Theorem 6.6 only admits improvement in finding the precise value of $C_2(n)$.

This statement can be applied to show that certain small sets are removable. For the meaning of $L^n \log^{n-1-C_2\beta} L$-capacity zero see the proof below.

Corollary 6.7. *Let* $\beta > 0$. *Let* $E \subset \mathbf{R}^n$ *be a closed set of* $L^n \log^{n-1-C_2\beta}$ L-*capacity zero where* C_2 *is the constant from Theorem 6.6 and let* $f : \Omega \setminus E \to \mathbf{R}^n$ *be a bounded mapping of finite distortion and assume that the distortion function* K *satisfies*

$$\int_{\Omega \setminus E} \exp(\beta K_f(x)) \, dx < \infty.$$

Then f *extends to a mapping of finite distortion in* Ω *with the exponentially integrable distortion function* $K_f(x)$.

Proof (of Theorem 6.6). We will use Lipschitz approximations similarly to the proof of Theorem 2.12. Let $\varphi \in C_0^\infty(B_0)$, $B_0 = B(x_0, r) \subset\subset \Omega$, and $\varphi \geq 0$. Let $u = f_i\varphi$ and extend it to be zero in $\mathbf{R}^n \setminus B_0$. Then $u \in W^{1,q}(\mathbf{R}^n)$ for all $q < n$, by the assumption on f in the theorem. Denote for $\lambda > 0$,

$$F_\lambda = \{x \in B(x_0, r) : M(g)(x) \leq \lambda \text{ and } x \text{ is a Lebesgue point of } u\},$$

where $g = |\varphi Df| + |f \otimes \nabla\varphi|$ in B_0 and $g = 0$ in $\mathbf{R}^n \setminus B_0$. Recall that $|f \otimes \nabla\varphi(x)| \approx \max_{i,j} |f_i(x)| |\frac{\partial\varphi(x)}{\partial x_j}|$ and hence clearly $|Du| \leq Cg$.

It is easy to show that u is $C\lambda$-Lipschitz continuous on the set $F_\lambda \cup (\mathbf{R}^n \setminus B_0)$ for $C = C(n) \geq 1$. Indeed, for $x, y \in F_\lambda$ we know this estimate from Lemma A.25. If $x \in F_\lambda$ and $y \in \mathbf{R}^n \setminus B_0$, set $\rho = 2\,\mathrm{dist}(x, \mathbf{R}^n \setminus B(x_0, r))$. Denote $A := \{x \in B(x, \rho) : u(x) = 0\}$. Since

$$|A| \geq |B(x,\rho) \cap (\mathbf{R}^n \setminus B_0)| \geq C(n)|B(x,\rho)|,$$

the Poincaré inequality yields

$$|u_{B(x,\rho)}| = \left| \fint_A (u - u_{B(x,\rho)}) \right| \leq \fint_A |u - u_{B(x,\rho)}| \leq C \fint_{B(x,\rho)} |u - u_{B(x,\rho)}|$$

$$\leq C(n)\rho \fint_{B(x,\rho)} |\nabla u| \leq C\rho Mg(x) \leq c\lambda|x - y|.$$

Thus analogously to the proof of Lemma A.25 we obtain

$$|u(x) - u(y)| = |u(x)| \leq |u(x) - u_{B(x,\rho)}| + |u_{B(x,\rho)}|$$
$$\leq C\rho M(|\nabla u|)(x) + C\lambda|x - y|$$
$$\leq C\rho Mg(x) + c\lambda|x - y| \leq C\lambda|x - y|.$$

If $x, y \in \mathbf{R}^n \setminus B_0$, then the claim is clear. Since all the other cases follow by symmetry, it follows that $u|_{F_\lambda \cup (\mathbf{R}^n \setminus B_0)}$ is $C\lambda$-Lipschitz continuous.

We extend $u|_{F_\lambda \cup (\mathbf{R}^n \setminus B_0)}$ to a Lipschitz continuous function u_λ on $\mathbf{R}^n$ with the same constant by the classical McShane extension theorem. Then we consider the mapping $f_\lambda = (u_\lambda, \varphi f_2, \varphi f_3, \ldots, \varphi f_n)$. Since $f \in W_{\mathrm{loc}}^{1,q}(\Omega, \mathbf{R}^n)$ for all $q < n$ and u_λ is Lipschitz we may apply Lemma 2.13 to obtain

$$\int_{B_0} J_{f_\lambda}(x)\,dx = 0,$$

and hence,

$$\int_{F_\lambda} J_{\varphi f}(x)\,dx = \int_{F_\lambda} J_{f_\lambda}(x)\,dx \leq -\int_{B_0 \setminus F_\lambda} J_{f_\lambda}(x)\,dx. \tag{6.16}$$

The estimate $|\nabla(\varphi f_i)| \leq C(n)g$ implies $|J_{f_\lambda}| \leq C(n)\lambda g^{n-1}$ and moreover we have $|f_i \nabla\varphi| \leq C(n)|f \otimes \nabla\varphi|$. Putting these estimates together and using the chain rule on the left hand-side of (6.16), we obtain that

$$\int_{F_\lambda} \varphi^n J_f(x)\,dx \leq C(n)\left(\int_{F_\lambda} |f \otimes \nabla\varphi| g^{n-1}\,dx + \lambda \int_{B_0 \setminus F_\lambda} g^{n-1}\,dx \right). \tag{6.17}$$

We claim that

$$\int_{\{g\le\lambda\}} \varphi^n J_f(x)\, dx \le C(n) \int_{\{g\le 2\lambda\}} |f \otimes \nabla\varphi| g^{n-1}\, dx + C(n)\lambda \int_{\{g>\lambda\}} g^{n-1}\, dx.$$
(6.18)

Indeed, by (6.2),

$$\int_{B_0\setminus F_\lambda} g^{n-1}\, dx \le \int_{\{g>\lambda\}} g^{n-1}\, dx + \lambda^{n-1} |\{x \in \mathbf{R}^n : Mg(x) > \lambda\}|$$

$$\le \int_{\{g>\lambda\}} g^{n-1}\, dx + C\lambda^{n-2} \int_{\{g>\lambda/2\}} g\, dx$$
(6.19)

$$\le C(n) \int_{\{g>\lambda/2\}} g^{n-1}\, dx,$$

and

$$\int_{\{g\le\lambda\}} \varphi^n J_f(x)\, dx \le \int_{\{Mg\le\lambda\}} \varphi^n J_f(x)\, dx + \lambda^n |\{x \in \mathbf{R}^n : Mg(x) > \lambda\}|$$
(6.20)

$$\le \int_{\{Mg\le\lambda\}} \varphi^n J_f(x)\, dx + C(n)\lambda \int_{\{g>\lambda/2\}} g^{n-1}\, dx.$$

Applying (6.20), (6.17), (6.19) and that $F_\lambda = \{Mg \le \lambda\} \subset \{g \le \lambda\}$ up to null sets, we obtain that

$$\int_{\{g\le\lambda\}} \varphi^n J_f(x)\, dx \le C(n) \int_{\{g\le\lambda\}} |f \otimes \nabla\varphi| g^{n-1}\, dx + C(n)\lambda \int_{\{g>\lambda/2\}} g^{n-1}\, dx.$$

Then (6.18) follows by replacing $\lambda/2$ by λ.

Now let $\alpha > 0$ be a constant, which will be chosen later. Note that

$$\Phi(\lambda) = \frac{1}{\lambda} \left(\log^{-(1+\alpha)} \lambda - (1 + \alpha) \log^{-(2+\alpha)} \lambda \right) \ge 0$$

for $\lambda \ge e^{1+\alpha}$. We multiply both sides of (6.18) by $\Phi(\lambda)$, and integrate with respect to λ over (t, ∞) for $t \ge \lambda_0 = \max\{e^{1+\alpha}, e^{2\alpha}\}$, and finally change the order of the integration to obtain that

$$\int_{B_0} \varphi^n J_f(x) \int_{\max\{g,t\}}^\infty \Phi(\lambda)\, d\lambda dx$$

$$\le C(n) \int_{B_0} |f \otimes \nabla\varphi| g^{n-1} \int_{\max\{g/2,t\}}^\infty \Phi(\lambda)\, d\lambda dx$$
(6.21)

$$+ C(n) \int_{\{g>t\}} g^{n-1} \int_t^g \lambda\Phi(\lambda)\, d\lambda dx.$$

Thus

$$\frac{1}{2\alpha}\int_{\{g<t\}}\frac{\varphi^n J_f(x)}{\log^\alpha t} + \frac{1}{2\alpha}\int_{\{g>t\}}\frac{\varphi^n J_f(x)}{\log^\alpha g}\,dx$$

$$\leq \int_{B_0}\varphi^n J_f(x)\left(\frac{1}{\alpha\log^\alpha\max\{g,t\}} - \frac{1}{\log^{1+\alpha}\max\{g,t\}}\right)dx \qquad (6.22)$$

$$\leq \frac{C(n)}{\alpha}\int_{B_0}\frac{|f\otimes\nabla\varphi|g^{n-1}}{\log^\alpha\max\{g/2,t\}}\,dx + C(n)\int_{\{g>t\}}\frac{g^n}{\log^{1+\alpha}g}\,dx.$$

Observe that the first inequality holds because $t \geq e^{2\alpha}$. We remark here that the assumption on the regularity of f in the theorem,

$$\frac{|Df|^n}{\log^{1+\alpha}(e+|Df|)} \in L^1_{loc}(\Omega),$$

implies that the integrals on the right-hand side of (6.22) are finite, where for the first term we need to use the Sobolev embedding Theorem A.18 similarly to Remark 2.11 (a).

Now we use the distortion inequality, which so far has not been used. The distortion inequality

$$|Df(x)|^n \leq K(x)J_f(x)$$

and the variant

$$ab \leq a\log(1+a) + e^b - 1$$

of Jensen's inequality for $a,b \geq 0$, imply that, in the set where $g(x) \geq \lambda_0$, we have

$$\frac{\varphi^n|Df|^n}{\log^{1+\alpha}g} \leq \frac{2}{\beta}\left(\frac{\varphi^n J_f}{\log^{1+\alpha}g}\right)\left(\frac{\beta}{2}K\right)$$

$$\leq \frac{2}{\beta}\left(\frac{3n\varphi^n J_f}{\log^\alpha g} + \exp\left(\frac{\beta}{2}K\right)\right). \qquad (6.23)$$

Therefore, recalling the definition of g and using (6.23),

$$\int_{\{g>t\}}\frac{g^n}{\log^{1+\alpha}g}\,dx \leq 2^n\int_{\{g>t\}}\frac{\varphi^n|Df|^n}{\log^{1+\alpha}g}\,dx + 2^n\int_{\{g>t\}}\frac{|f\otimes\nabla\varphi|^n}{\log^{1+\alpha}g}\,dx$$

$$\leq \frac{C(n)}{\beta}\int_{\{g>t\}}\frac{\varphi^n J_f}{\log^\alpha g}\,dx + \frac{C(n)}{\beta}\int_{\{g>t\}}\exp\left(\frac{\beta}{2}K\right)dx \qquad (6.24)$$

$$+ C(n)\int_{\{g>t\}}\frac{|f\otimes\nabla\varphi|^n}{\log^{1+\alpha}g}\,dx.$$

Inserting (6.24) in (6.22), and rearranging, it follows that

$$
\frac{1}{2\alpha \log^\alpha t} \int_{\{g<t\}} \varphi^n J_f
$$

$$
\leq \left(\frac{C(n)}{\beta} - \frac{1}{2\alpha}\right) \int_{\{g>t\}} \frac{\varphi^n J_f}{\log^\alpha g}\, dx + \frac{C(n)}{\beta} \int_{\{g>t\}} \exp\left(\frac{\beta}{2}K\right) dx \qquad (6.25)
$$

$$
+ C(n) \int_{\{g>t\}} \frac{|f \otimes \nabla\varphi|^n}{\log^{1+\alpha} g}\, dx + \frac{C(n)}{\alpha} \int_{B_0} \frac{|f \otimes \nabla\varphi| g^{n-1}}{\log^\alpha \max\{g/2, t\}}\, dx\, .
$$

Now let us fix $\alpha = \beta/4C(n)$ to be the constant in the theorem. We multiply both sides of (6.25) by $\log^\alpha t$ and let $t \to \infty$. We obtain by monotone convergence theorem and Lebesgue dominated convergence theorem that

$$
\int_{B_0} \varphi^n J_f\, dx \leq C(n) \int_{B_0} |f \otimes \nabla\varphi| g^{n-1}\, dx. \qquad (6.26)
$$

Here we used the integrability of $|f \otimes \nabla\varphi| g^{n-1}, |f \otimes \nabla\varphi|^n$, and $\log g > \log t$ on the set $\{g > t\}$ to deal with the first and third term of the right-hand side of (6.25). The second term can be estimated using Hölder's inequality, the integrability of $\exp(\beta K(x))$ and an elementary estimate on the measure of the set $\{g > t\}$. Hence, (6.26) shows that $J_f(x) \in L^1_{\mathrm{loc}}(\Omega)$, and then $|Df| \in L^n \log^{-1} L_{\mathrm{loc}}$ follows from Lemma 2.8. $\qquad \square$

Proof (of Corollary 6.7). First we prove the following Cac) Cacioppoli type inequality:

$$
\int_{B_0} \frac{|Df|^n \varphi^n}{\log^{1+\alpha}(e + |Df|\varphi)}\, dx \leq C(n,\beta) \int_{B_0} \frac{|f \otimes \nabla\varphi|^n}{\log^{\alpha+1-n}(e + |f \otimes \nabla\varphi|)}\, dx
$$

$$
+ C(n,\beta) \int_{B_0} \exp\left(\frac{\beta}{2} K(x)\right) dx. \qquad (6.27)
$$

To this end, we insert (6.22) into (6.24). We obtain that

$$
\int_{\{g>t\}} \frac{g^n}{\log^{1+\alpha}(e + g)}\, dx
$$

$$
\leq C(n,\beta)\left[\int_{B_0} \frac{|f \otimes \nabla\varphi| g^{n-1}}{\log^\alpha \max(g/2, t)}\, dx + \int_{\{g>t\}} \frac{|f \otimes \nabla\varphi|^n}{\log^{1+\alpha}(e + g)}\, dx \quad (6.28)\right.
$$

$$
\left. + \int_{\{g>t\}} \exp\left(\frac{\beta}{2} K(x)\right) dx\right] + \frac{2\alpha C(n)}{\beta} \int_{\{g>t\}} \frac{g^n}{\log^{1+\alpha}(e + g)}\, .
$$

By the choice of α made before, the last term in the right-hand side may be absorbed to the left and therefore ignored. Thus, letting $t = \lambda_0$ in (6.28) and noticing that $\exp(\frac{\beta}{2}K(x)) \geq 1$, results in

$$\int_{B_0} \frac{g^n}{\log^{1+\alpha}(e+g)}\, dx \le C(n,\beta)\left[\int_{B_0} \frac{|f \otimes \nabla\varphi|g^{n-1}}{\log^\alpha(e+g)}\, dx \right. \tag{6.29}$$

$$\left. + \int_{B_0} \frac{|f \otimes \nabla\varphi|^n}{\log^{1+\alpha}(e+g)}\, dx + \int_{B_0} \exp(\frac{\beta}{2}K(x))\, dx \right].$$

To estimate the first integral in the right-hand side of (6.29), we use the inequality

$$ab^{n-1} \le \varepsilon \frac{b^n}{\log(e+b)} + C(\varepsilon,n)a^n \log^{n-1}(e+a)$$

for non-negative numbers a and b, and obtain that

$$\frac{|f \otimes \nabla\varphi|g^{n-1}}{\log^\alpha(e+g)} \le \frac{|f \otimes \nabla\varphi|}{\log^{\frac{\alpha}{n}}(e+|f \otimes \nabla\varphi|)} \cdot \frac{g^{n-1}}{\log^{\alpha\frac{n-1}{n}}(e+g)}$$

$$\le \varepsilon C(\alpha,n)\frac{g^n}{\log^{1+\alpha}(e+g)} + C(\varepsilon,n)\frac{|f \otimes \nabla\varphi|^n}{\log^{\alpha+1-n}(e+|f \otimes \nabla\varphi|)}.$$

By taking $\varepsilon = C(n,\beta)/2C(\alpha,n)$, where $C(n,\beta)$ is the constant in (6.29), we infer from the definition of g and (6.29) that

$$\int_{B_0} \frac{|Df|^n\varphi^n}{\log^{1+\alpha}(e+|Df|\varphi)}\, dx \le \int_{B_0} \frac{g^n}{\log^{1+\alpha}(e+g)}\, dx$$

$$\le C(n,\beta)\int_{B_0} \frac{|f \otimes \nabla\varphi|^n}{\log^{\alpha+1-n}(e+|f \otimes \nabla\varphi|)}\, dx$$

$$+C(n,\beta)\int_{B_0} \frac{|f \otimes \nabla\varphi|^n}{\log^{\alpha+1}(e+|f \otimes \nabla\varphi|)}\, dx$$

$$+C(n,\beta)\int_{B_0} \exp(\frac{\beta}{2}K(x))\, dx$$

$$\le C(n,\beta)\int_{B_0} \frac{|f \otimes \nabla\varphi|^n}{\log^{\alpha+1-n}(e+|f \otimes \nabla\varphi|)}\, dx$$

$$+C(n,\beta)\int_{B_0} \exp(\frac{\beta}{2}K(x))\, dx,$$

which proves (6.27).

We note that E has vanishing $(n-1)$-dimensional Hausdorff measure, in fact it has Hausdorff dimension zero. Therefore it is easy to see using the ACL-condition that $f \in W^{1,1}(\Omega, \mathbf{R}^n)$ and that it satisfies the distortion inequality almost everywhere. Now we need to verify the assumptions of Theorem 6.6, i.e. we need to show that

$$\frac{|Df|^n}{\log^{1+\alpha}(e+|Df|)} \in L^1_{\text{loc}}(\Omega), \tag{6.30}$$

where $\alpha = C_2\beta$ is as in Theorem 6.6. To this end, let $\eta \in C_0^\infty(\Omega)$ be an arbitrary nonnegative test function. We denote by E' the intersection of E with the support of η. There exists a sequence of functions $\{\phi_j\}_{j=1}^\infty$ such that for each j we have

1. $\phi_j \in C_0^\infty(\Omega)$,

2. $0 \le \phi_j \le 1$,

3. $\phi_j = 1$ on some neighborhood U_j of E',

4. $\lim_{j\to\infty} \phi_j(x) = 0$ for almost all $x \in \mathbf{R}^n$,

5. $\lim_{j\to\infty} \int_\Omega |\nabla\phi_j|^n \log^{n-1-\alpha}(e+|\nabla\phi_j|) = 0$.

We set

$$\varphi_j = (1-\phi_j)\eta \in C_0^\infty(\text{spt}\,\eta \setminus E').$$

Let $\varphi = \varphi_j$ in (6.27). Recall that $|f|$ is assumed to be bounded in Ω and also that $\nabla\varphi_j = (1-\phi_j)\nabla\eta - \eta\nabla\phi_j$. It follows from the conditions defining ϕ_j that we can pass to the limit (as $j \to \infty$) in (6.27) to obtain that

$$\int_{B_0} \frac{|Df|^n \eta^n}{\log^{1+\alpha}(e+|Df|^n\eta^n)}\,dx \le C(n,\beta)\int_{B_0}\frac{|f\otimes\nabla\eta|^n}{\log^{\alpha+1-n}(e+|f\otimes\nabla\eta|)}\,dx$$
$$+ C(n,\beta)\int_{B_0}\exp\left(\frac{\beta}{2}K(x)\right)dx,$$

which implies (6.30). By Theorem 6.6 we obtain $J_f \in L^1_{\text{loc}}$ which shows that f is a mapping of finite distortion and Corollary 6.7 follows. □

Remark 6.8. The results of this section were established by Faraco et al. in [30]. For earlier related results see [3].

Open problem 19. What is the optimal value of C_2 in Theorem 6.6 (or in Corollary 6.7)? An example from [3] shows that the capacity considered in Corollary 6.7 is of the correct form.

Chapter 7
Final Comments

Abstract In this chapter we briefly discuss the inner distortion function. We also give the connection in the plane between mappings of finite distortion and solutions to a degenerate Beltrami equation. Finally, we study the shape of the image of the unit disk under a mapping of finite distortion and we show that certain families of mappings with exponentially integrable distortion are closed under weak convergence.

7.1 Inner Distortion

For mappings of finite distortion we have defined the distortion function K_f. It is often called the outer distortion function and referred to by

$$K_O(x) := \begin{cases} \frac{|Df(x)|^n}{J_f(x)} & \text{for } J_f(x) > 0 \, , \\ 1 & \text{for } J_f(x) = 0 \, . \end{cases}$$

It is possible to define also other distortion functions such as the inner distortion function

$$K_I(x) := \begin{cases} \frac{|\operatorname{adj} Df(x)|^n}{J_f(x)^{n-1}} & \text{for } J_f(x) > 0 \, , \\ 1 & \text{for } J_f(x) = 0 \, , \end{cases}$$

where $\operatorname{adj} Df(x)$ denotes the adjugate matrix of $Df(x)$, i.e. the matrix of the $(n-1) \times (n-1)$ subdeterminants. These distortion functions coincide for $n = 2$ but they are different for $n \geq 3$.

We have the following geometrical interpretation. Let E be the ellipsoid defined as $E = \{Df(x)z \in \mathbf{R}^n : |z| \leq 1\}$. Then K_O corresponds (modulo a dimensional constant) to the ratio of the longest axis of E to power n divided by the volume

S. Hencl and P. Koskela, *Lectures on Mappings of Finite Distortion*, Lecture Notes in Mathematics 2096, DOI 10.1007/978-3-319-03173-6_7,
© Springer International Publishing Switzerland 2014

of E and K_I corresponds (modulo a dimensional constant) to the ratio of the $(n-1)$-dimensional volume of the largest intersection of E with $(n-1)$-dimensional hyperplane to power n divided by the volume of E to power $n-1$. Roughly speaking the outer distortion corresponds to the deformation of lengths of segments and the inner distortion corresponds to the deformation of the $(n-1)$-dimensional volumes of intersection with hyperplanes.

Let $Df(x)$ be a diagonal matrix with strictly positive entries

$$\lambda_1 \geq \lambda_2 \geq \ldots \geq \lambda_n > 0 .$$

Then

$$|Df(x)| = \lambda_1, \ |\operatorname{adj} Df(x)| = \lambda_1 \lambda_2 \cdots \lambda_{n-1} \text{ and } J_f(x) = \lambda_1 \lambda_2 \cdots \lambda_n .$$

Hence it is easy to see that $|\operatorname{adj} Df(x)| \leq |Df(x)|^{n-1}$ and therefore

$$K_I(x) = \frac{|\operatorname{adj} Df(x)|^n}{J_f(x)^{n-1}} \leq \frac{|Df(x)|^{n(n-1)}}{J_f(x)^{n-1}} \leq K_O(x)^{n-1} .$$

Moreover, we have an equality if $\lambda_1 = \lambda_2 = \ldots = \lambda_{n-1}$. We may also estimate

$$|Df(x)| = \lambda_1 \leq \lambda_1 \frac{\lambda_2}{\lambda_n} \cdots \frac{\lambda_{n-1}}{\lambda_n} = \frac{(\lambda_1 \cdots \lambda_{n-1})^{n-1}}{(\lambda_1 \cdots \lambda_n)^{n-2}} = \frac{|\operatorname{adj} Df(x)|^{n-1}}{J_f(x)^{n-2}}$$

It follows that

$$K_O(x) = \frac{|Df(x)|^n}{J_f(x)} \leq \frac{|\operatorname{adj} Df(x)|^{n(n-1)}}{J_f(x)^{(n-1)(n-1)}} = K_I(x)^{n-1}$$

and for $\lambda_2 = \lambda_3 = \ldots = \lambda_n$ we have in fact an equality. This shows that the double inequality

$$K_I^{\frac{1}{n-1}}(x) \leq K_O(x) \leq K_I^{n-1}(x) \tag{7.1}$$

that can be proven for a general matrix with positive determinant by linear algebra is sharp.

Similarly to the proof of Corollary 1.9 we can use the proof of Theorem 5.9 to establish the following identity.

Corollary 7.1. *Let $\Omega \subset \mathbf{R}^n$ be a domain and let $f \in W^{1,n-1}(\Omega, f(\Omega))$ be a homeomorphism of finite distortion with $K_I \in L^1(\Omega)$. Then $f^{-1} \in W^{1,n}_{\text{loc}}(f(\Omega), \mathbf{R}^n)$ is a mapping of finite distortion and*

$$\int_\Omega K_I(x) \, dx = \int_{f(\Omega)} |Df^{-1}(y)|^n \, dy .$$

From inequality (7.1) and Theorem 2.4 we may deduce that for continuity and other properties of mappings of finite (outer) distortion it is enough to assume that $\exp(\lambda K_I^{n-1}) \in L^1_{loc}(\Omega)$. However already a much weaker condition suffices.

Theorem 7.2. *Let $\Omega \subset \mathbf{R}^n$ be open and let $f : \Omega \to \mathbf{R}^n$ be a mapping of finite distortion. Suppose that there is $\lambda > (n-1)^2 - 1$ such that $\exp(\lambda \sqrt[n-1]{K_I}) \in L^1_{loc}(\Omega)$. Then f is continuous.*

Remark 7.3. (a) It is possible to define other distortion functions corresponding to intersections with k-dimensional hyperplanes and we recommend [67, Chap. 6.5] for further reading.

(b) Theorem 7.2 was proven by Onninen in [100]. The proof uses the fact that, under these assumptions, the distributional Jacobian coincides with the usual Jacobian which was shown by Giannetti et al. in [36].

(c) For $n \geq 3$ the example $f(x) = (u(x), 0, \ldots, 0)$ with discontinuous u shows that for continuity in Theorem 7.2 we cannot replace the assumption of finite outer distortion $(J_f(x) = 0 \Rightarrow |Df(x)| = 0)$ by the condition $(J_f(x) = 0 \Rightarrow |\operatorname{adj} Df(x)| = 0)$.

Open problem 20. Does Theorem 7.2 hold without the artificial looking assumption $\lambda > (n-1)^2 - 1$?

Open problem 21. Let $g : \overline{B(0,1)} \to \overline{B(0,1)}$ be a homeomorphism of finite distortion with $K_I \in L^1(B(0,1))$, where $p \geq 1$. Consider the minimization problem of $\int_{B(0,1)} K_I$ for mappings $f : B(0,1) \to \mathbf{R}^n$ of finite distortion that coincide on the boundary of $B(0,1)$ with g. Does there exist a diffeomorphic minimizer? Here K_I is defined using the Hilbert-Schmidt norm. For $p = 1$, the identity in Corollary 7.1 also holds for this distortion and the Hilbert-Schmidt norm of Df^{-1}.

The following problem is open even in the planar setting.

Open problem 22. Given a bounded domain Ω and a homeomorphism $g : \overline{\Omega} \to \overline{f(\Omega)}$ of finite distortion with $K_I \in L^p(B(0,1))$, where $p > 1$, is every homeomorphism f of finite distortion that minimizes $\int_{B(0,1)} K_I^p$ with the given boundary values necessarily a smooth diffeomorphism?

7.2 Ball's Question from the Introduction

Let us now show that our results imply the positive answer to the question from the introduction. We give the full proof in the planar case but in higher dimensions we need the following theorem from [112]. Its proof requires the concept of modulus of a surface family that we have decided not to discuss.

Theorem 7.4. *Let $\Omega \subset \mathbf{R}^n$ be a domain and let $f \in W^{1,n}(\Omega, \mathbf{R}^n)$ be a mapping of finite distortion such that $K_I \in L^1(\Omega)$. Further we assume that there is compact* set $E \subset \Omega$ such that f is a homeomorphism of $\Omega \setminus E$ onto $f(\Omega) \setminus f(E)$. Then f is open and discrete.

Theorem 7.5. *Let $f : B(0, 1) \to B(0, 1)$ be a mapping in $\mathbf{R}^n$ for some $n \geq 2$ such that f is a homeomorphism from $B(0, 1) \setminus \overline{B(0, 1 - \delta)}$ onto $B(0, 1) \setminus f(\overline{B(0, 1 - \delta)})$, $f \in W^{1,n}(B(0, 1), \mathbf{R}^n)$, $J_f(x) > 0$ almost everywhere and*

$$\int_{B(0,1)} |(Df(x))^{-1}|^n J_f(x) \, dx < \infty. \tag{7.2}$$

Then there is a continuous representative of f, it is a homeomorphism in $B(0, 1)$ and its inverse satisfies $f^{-1} \in W^{1,n}(B(0, 1), \mathbf{R}^n)$.

Proof. From $f \in W^{1,n}$ we obtain $J_f \in L^1$ and hence the condition $J_f(x) > 0$ clearly implies that f is a mapping of finite distortion. From Theorem 2.3 we know that f has a continuous representative.

By simple linear algebra (A.1) we know that the adjugate matrix satisfies

$$Df(x) \operatorname{adj} Df(x) = I J_f(x)$$

and hence for every x such that $J_f(x) > 0$ we obtain

$$K_I(x) = \frac{|\operatorname{adj} Df(x)|^n}{J_f(x)^{n-1}} = \frac{|(Df(x))^{-1} J_f(x)|^n}{J_f(x)^{n-1}} = |(Df(x))^{-1}|^n J_f(x) \,.$$

Our assumption (7.2) thus implies that $K_I \in L^1$.

Let us first finish the proof in the planar case. For $n = 2$ we know that $K_f(x) = K_I(x) \in L^1$. Theorem 3.4 now shows that f is an open and discrete mapping and by Theorem 3.27 we get that f is a homeomorphism on the whole $B(0, 1)$. Finally Theorem 5.9 shows that $f^{-1} \in W^{1,n}$.

For $n \geq 3$ Theorem 7.4 guarantees that f is both open and discrete. Analogously to the proof of Theorem 3.27 we can thus obtain that f is a homeomorphism on the whole $B(0, 1)$. The conclusion that $f^{-1} \in W^{1,n}$ follows from the proof of Theorem 5.9, see second line of (5.8) there.   □

Under higher integrability of the distortion we can obtain a similar result for mappings that are not necessarily homeomorphisms close to the boundary but only equal to a homeomorphism on the boundary.

Remark 7.6. Let $f : \overline{B(0, 1)} \to \mathbf{R}^n$, $n \geq 2$, be a continuous mapping such that $f = g$ on $\partial B(0, 1)$ for some homeomorphism $g : \overline{B(0, 1)} \to \mathbf{R}^n$. Suppose that $f \in W^{1,n}(B(0, 1), \mathbf{R}^n)$ is a mapping of finite distortion such that $K_f \in L^p$ for some $p > n - 1$ if $n \geq 3$ and $p = 1$ for $n = 2$. Then f is a homeomorphism in $B(0, 1)$ and its inverse satisfies $f^{-1} \in W^{1,n}(B(0, 1), \mathbf{R}^n)$. For related results see [118].

Sketch of the Proof. From Remark 3.12 (a) we know that f and g have the same degree because they agree on the boundary. The degree of a homeomorphism on a domain is either identically $+1$ or -1 (see [117, p. 17] or [108]). Since f is differentiable a.e. by Theorem 2.24 and $J_f > 0$ a.e. by Theorem 4.13 we obtain that the degree of f equals to $+1$ by Remark 3.17.

We claim that each point $y \in f(B(0,1))$ has at most one preimage. Suppose on the contrary that we have two points $x_1, x_2 \in f^{-1}(y)$. By Theorem 3.4 we know that f is open and discrete and from its proof we know that $\mathcal{H}^1(f^{-1}(y)) = 0$. We can proceed similarly to the proof of Theorem 3.18 and we can choose neighborhoods U_1, U_2 such that $x_1 \in U_1$, $x_2 \in U_2$ and

$$y \notin f(\partial U_1), \ y \notin f(\partial U_2) \text{ and } U_1 \cap U_2 = \emptyset .$$

By openness we know that the set $f(U_1) \cap f(U_2)$ is open and hence we can find a smooth nonnegative function $\varphi \in C_0^\infty(C \cap f(U_1) \cap f(U_2))$ such that $\int \varphi = 1$. Hence we may find compact sets $E_i \subset U_i$ with $J_f > 0$ (recall that $J_f > 0$ a.e.) on each E_i and we may moreover assume that $\varphi > 0$ on $f(E_1) \cup f(E_2)$. By Theorem 3.15, the nonnegativity of the Jacobian and Theorem 3.16 we obtain

$$\deg(y, f, B(0,1)) = 1 = \int_{B(0,1)} \varphi(f(x)) J_f(x) \, dx$$

$$\geq \sum_{i=1}^{2} \int_{U_i} \varphi(f(x)) J_f(x) \, dx = \sum_{i=1}^{2} \deg(y, f, U_i) \geq 2$$

which gives us the desired contradiction. Above we have used Theorem 3.15 for the entire domain $B(0,1)$ which is not in principle allowed. However, f is continuous up to the boundary and hence the proof of Theorem 3.15 allows us to do that.

It follows that f is a homeomorphism in $B(0,1)$. Finally Theorem 5.9 shows that $f^{-1} \in W^{1,n}$. $\qquad \square$

7.3 Beltrami Equations

In this section we briefly explain the connection of mappings of bounded and finite distortion in the plane with the solutions of Beltrami equations. It is convenient to use complex notation: we identify $\mathbf{R}^2$ with $\mathbf{C}$ and write a point $z \in \mathbf{C}$ as $z = x + iy$, where x, y are real. Let $f \in W^{1,1}_{\text{loc}}(\Omega; \mathbf{C})$ be continuous, where $\Omega \subset C$ is a domain. Writing $f(z) = u(z) + iv(z)$ with u, v real-valued, we notice that both u and v have, at almost every z, partial derivatives u_x, u_y, v_x, v_y with respect to x, y. Then

$$\partial_x f(z) = u_x(z) + iv_x(z),$$
$$\partial_y f(z) = u_y(z) + iv_y(z).$$

We will employ the derivatives $\partial f, \bar{\partial} f$ defined by

$$\partial f(z) = \frac{1}{2}(\partial_x f(z) - i\partial_y f(z)),$$

$$\bar{\partial} f(z) = \frac{1}{2}(\partial_x f(z) + i\partial_y f(z)).$$

Recalling the Cauchy-Riemann equations

$$u_x = v_y, u_y = -v_x,$$

we notice that $\bar{\partial} f(z) = 0$ if f is analytic. In fact, for a continuous $f \in W_{loc}^{1,1}(\Omega; \mathbf{C})$, $\bar{\partial} f(z) = 0$ almost everywhere only when f is analytic.

Let us further denote by $\partial_\alpha f(z)$ the derivative of f in the direction $e^{i\alpha}$ (if it happens to exist). In the real notation, this is simply $Df(x, y)(\cos\alpha, \sin\alpha)$ if f is differentiable at the point (x, y) and it is easy to check that, in our complex notation,

$$\partial_\alpha f(z) = \partial f(z)e^{i\alpha} + \bar{\partial} f(z)e^{-i\alpha}. \tag{7.3}$$

In fact, one has for each $h \in \mathbf{C}$

$$Df(z)h = \partial f(z)h + \bar{\partial} f(z)\bar{h},$$

where $\bar{h}$ is the complex conjugate of h (for $h = x + iy$, $\bar{h} = x - iy$). Now $\partial_\alpha f(z)$ has maximal length when the two vectors in the sum (7.3) point to the same direction, i.e. when

$$\alpha + \arg \partial f(z) = -\alpha + \arg \bar{\partial} f(z)$$

(modulo 2π), and minimal length when these two vectors point to opposite directions. Here $\arg w$ denotes the argument of a complex number w. Thus the maximal directional derivative has the value

$$|\partial f(z)| + |\bar{\partial} f(z)|$$

and corresponds to the choice

$$\alpha = \frac{1}{2}(\arg \bar{\partial} f(z) - \arg \partial f(z))$$

and one has the minimal value

$$||\partial f(z)| - |\bar{\partial} f(z)||$$

corresponding to

$$\alpha = \frac{\pi}{2} + \frac{1}{2}(\arg \overline{\partial} f(z) - \arg \partial f(z)).$$

Moreover,

$$|J_f(z)| = |(|\partial f(z)| + |\overline{\partial} f(z)|)(|\partial f(z)| - |\overline{\partial} f(z)|)|$$
$$= ||\partial f(z)|^2 - |\overline{\partial} f(z)|^2|.$$

Theorem 7.7. *Let* $\mu : \mathbf{C} \to \mathbf{C}$ *satisfy* $||\mu||_{L^\infty} < 1$. *Then there is a quasiconformal mapping* $f : \mathbf{C} \to \mathbf{C}$ *so that*

$$\overline{\partial} f(z) = \mu(z)\partial f(z)$$

almost everywhere.

This is a very strong existence theorem. Notice that $J_f(z) \neq 0$ almost everywhere because f is quasiconformal. Thus the discussion before Theorem 7.7 shows that

$$\frac{|Df(z)|^2}{|J_f(z)|} = \frac{||\partial f(z)| + |\overline{\partial} f(z)||^2}{||\partial f(z)|^2 - |\overline{\partial} f(z)|^2|} = \frac{1 + |\mu(z)|}{1 - |\mu(z)|}$$

almost everywhere. Moreover, for almost every z,

$$K_f(z) = \frac{1 + |\mu(z)|}{1 - |\mu(z)|}$$

and the differential $Df(z)$ maps disks $B(z, r)$ centered at z to ellipses with major axes of the length

$$2|Df(z)|r = 2|\partial f(z)|r(1 + |\mu(z)|)$$

and minor axes of the length

$$2|\partial f(z)|r(1 - |\mu(z)|).$$

The orientation of these ellipses is not determined by $\mu(z)$. However, consider the collection of all ellipses E with center z so that the ratio of the major and the minor axis is $K_f(z)$ and the angle determined by the minor axis and the real line is

$$\alpha = \frac{1}{2}\arg \mu(z).$$

Then the differential $Df(z)$ maps these ellipses to disks centered at $f(z)$.

We will omit the proof of Theorem 7.7 and refer the reader to [4] for the proof and further extensions of this existence theorem. The theory of singular integral operators is one of the main ingredients in its proof.

Let us recall the Riemann mapping theorem, see [103, p. 420] for a proof.

Theorem 7.8 (Riemann Mapping Theorem). *Each simply connected domain* $\Omega \subsetneq \mathbf{C}$ *is conformally equivalent to the unit disk.*

It follows that, given simply connected, proper subdomains Ω, Ω' of the plane, there is a conformal mapping $f : \Omega \to \Omega'$. We continue with a quasiconformal version of this statement.

Theorem 7.9 (Measurable Riemann Mapping Theorem). *Let* $\Omega, \Omega' \subsetneq \mathbf{C}$ *be simply connected subdomains and suppose that* $\mu : \Omega \to \mathbf{C}$ *satisfies* $\|\mu\|_{L^\infty} < 1$. *Then there is a quasiconformal mapping* $f : \Omega \to \Omega'$ *so that*

$$\overline{\partial} f(z) = \mu(z) \partial f(z) \quad a.e. \text{ in } \Omega. \tag{7.4}$$

In fact, f is $\dfrac{1 + \|\mu\|_\infty}{1 - \|\mu\|_\infty}$ *-qc.*

Proof. Given Ω, Ω' and μ, we extend μ as zero to the rest of $\mathbf{C}$. Then Theorem 7.7 gives us a quasiconformal mapping as asserted, except for the requirement that $f(\Omega) = \Omega'$. In any case, $f(\Omega)$ is a simply connected proper subdomain of $\mathbf{C}$, and thus the usual Riemann mapping theorem provides us with a conformal mapping $g : f(\Omega) \to \Omega'$. Setting $\tilde{f} = g \circ f$, it is easy to check using the "chain rules"

$$\partial(g \circ f) = \partial g(f) \partial f + \overline{\partial} g(f) \partial \overline{f},$$
$$\overline{\partial}(g \circ f) = \partial g(f) \overline{\partial} f + \overline{\partial} g(f) \overline{\partial \overline{f}},$$

that $\tilde{f}$ has all the required properties. By multiplying the first equation by $\mu(z)$ and using $\overline{\partial} f(z) = \mu(z) \partial f(z)$ we easily get the expressions in the second equation and thus also (7.4). $\qquad\qquad\qquad\qquad\qquad\qquad\qquad\qquad\qquad\qquad\qquad\qquad\qquad\square$

Moreover, there are more general existence results that assume that we have a mapping of exponentially integrable distortion, see [4, Theorem 20.4.9] for a proof.

Theorem 7.10. *Let* $\mu : \mathbf{C} \to \mathbf{C}$ *satisfy* $\mu(z) = 0$ *for* $|z| \geq 1$ *and suppose that*

$$K(z) = \frac{1 + |\mu(z)|}{1 - |\mu(z)|} \text{ satisfies } \exp(\lambda K) \in L^1(B(0, 1)) \text{ for some } \lambda > 0.$$

Then there is a homeomorphic solution $f \in WL^n \log^{-1} L_{\text{loc}}(\mathbf{C})$ *to the Beltrami equation* $\overline{\partial} f(z) = \mu(z) \partial f(z)$. *Moreover, every other* $WL^n \log^{-1} L$-solution h *to this Beltrami equation in a domain* $\Omega \subset \mathbf{C}$ *admits the factorization*

$$h = \phi \circ f$$

Fig. 7.1 Model domains Ω_s
and G_s

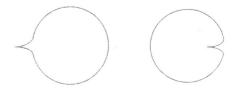

where ϕ is a holomorphic function in the domain $f(\Omega)$.

7.4 Shape of the Image of a Disk

The geometry of those planar Jordan domains Ω that arise as images of the
unit disk $B(0, 1)$ under a quasiconformal homeomorphism $f : \mathbf{R}^2 \to \mathbf{R}^2$
is precisely understood. Recall here that a quasiconformal mapping is, in our
terminology, a homeomorphic mapping of finite distortion with $K_f \in L^\infty(\mathbf{R}^n)$.
The characterizing property for a Jordan domain Ω to be the image of $B(0, 1)$ under
a quasiconformal mapping of the entire plane is the Ahlfors three point property:

$$\min_{i=1,2} \text{diam}(\gamma_i) \leq C |P_1 - P_2| \tag{7.5}$$

for any $P_1, P_2 \in \partial\Omega$ with $P_1 \neq P_2$, where γ_1, γ_2 are the components of
$\partial\Omega \setminus \{P_1, P_2\}$ and C is a constant that depends on Ω. This condition rules out
both exterior and interior cusps.

   We would like to have criteria to decide whether a given Jordan domain can arise
as the image of the unit disk under a global homeomorphism of finite distortion with
integrability restrictions on the distortion function.

   If we relax the condition $K_f \in L^\infty(\mathbf{R}^2)$ to $\exp(\lambda K_f) \in L^1_{\text{loc}}(\mathbf{R}^2)$ for some
$\lambda > 0$, both exterior and interior cusps are possible. Let us consider the model
domains Ω_s, G_s (see Fig. 7.1) defined for $s > 0$ by

$$\Omega_s = \{(x_1, x_2) \in \mathbf{R}^2 : 0 < x_1 < 1, \ |x_2| < x_1^{1+s}\} \cup B(x_s, r_s),$$

where $x_s = (s + 2, 0)$ and $r_s = \sqrt{(s + 1)^2 + 1}$, and

$$G_s = B(x'_s, r_s) \setminus \{(x_1, x_2) \in \mathbf{R}^2 : x_1 \geq 0, \ |x_2| \leq x_1^{1+s}\},$$

where $x'_s = (-s, 0)$.

   The following result for the model domains Ω_s can be found in [82]; the case of
G_s can easily be reduced to this via a suitable inversion.

Theorem 7.11. *Let $s > 0$. For each $\lambda < \frac{2}{s}$ there exists a homeomorphic mapping $f : \mathbf{R}^2 \to \mathbf{R}^2$ of finite distortion so that $\exp(\lambda K_f) \in L^1_{\mathrm{loc}}(\mathbf{R}^2)$ and $f(B(0,1)) = \Omega_s$. On the other hand, for $\lambda > \frac{2}{s}$, no such mapping exists.*

For $\lambda < \frac{2}{s}$ there is also a homeomorphic mapping $f : \mathbf{R}^2 \to \mathbf{R}^2$ of finite distortion so that $\exp(\lambda K_f) \in L^1_{\mathrm{loc}}(\mathbf{R}^2)$ and $f(B(0,1)) = G_s$. while no such a mapping exists for $\lambda > \frac{2}{s}$.

Since every simply connected planar domain is conformally equivalent to the unit disk, it is also natural to consider these model domains Ω_s, G_s in the setting where f is additionally required to be quasiconformal in the unit disk. The following result is from [80].

Theorem 7.12. *Let $s > 0$. For each $\lambda < \frac{1}{s}$ there is a homeomorphic mapping $f : \mathbf{R}^2 \to \mathbf{R}^2$ of finite distortion so that $\exp(\lambda K_f) \in L^1_{\mathrm{loc}}(\mathbf{R}^2)$, f is quasiconformal in $B(0,1)$ and $f(B(0,1)) = \Omega_s$. On the other hand, for $\lambda > \frac{1}{s}$, no such mapping exists.*

The situation for G_s is different, see [41].

Theorem 7.13. *Let $s > 0$. If $f : \mathbf{R}^2 \to \mathbf{R}^2$ is a homeomorphic mapping of finite distortion so that f is quasiconformal in $B(0,1)$ and $f(B(0,1)) = G_s$, then $K_f \notin L^p(B(0,2))$ if $ps > \|K_f\|_{L^\infty(B(0,1))}$.*

Notice that exterior and interior cusps are very different from the point of view of the degree of integrability of K_f under the additional quasiconformality requirement; compare with Theorem 7.11.

By combining recent results from [41] and [43] one obtains a sufficient version of the Ahlfors three point property.

Theorem 7.14. *Let $\Omega \subset \mathbf{R}^2$ be a Jordan domain. Suppose that there is a constant C so that*

$$\min_{i=1,2} \mathrm{diam}(\gamma_i) \leq C\varphi(|P_1 - P_2|)$$

for all $P_1, P_2 \in \partial\Omega$ with $P_1 \neq P_2$, where γ_1, γ_2 are the components of $\partial\Omega \setminus \{P_1, P_2\}$ and $\varphi(t) = t(\log\log\frac{1}{t})^{1/5}$. Then any quasiconformal mapping $f : B(0,1) \to \Omega$ extends to a homeomorphism $\hat{f} : \mathbf{R}^2 \to \mathbf{R}^2$ of finite distortion with $\exp(\lambda K_{\hat{f}}) \in L^1_{\mathrm{loc}}(\mathbf{R}^2)$ for some $\lambda > 0$.

The form of the function φ rules out both Ω_s and G_s. On the other hand, Theorem 7.13 shows that G_s should be ruled out, and Theorem 7.14 is not restricted to domains of model type.

Open problem 23. It seems plausible that the function φ from Theorem 7.14 is not optimal. On the other hand, from [41] we know that it cannot be replaced by $\hat{\varphi}(t) = t \log^{1+\varepsilon}(\frac{1}{t})$ for any $\varepsilon > 0$. How much can one relax φ?

For some partial results in this direction see Guo [42].

Open problem 24. How much can the function φ from Theorem 7.14 be relaxed if one only asks for $\Omega = f(B(0,1))$ for some homeomorphic mapping $f : \mathbf{R}^2 \to \mathbf{R}^2$ of finite distortion so that $\exp(\lambda K_f) \in L^1_{\text{loc}}(\mathbf{R}^2)$ for some $\lambda > 0$? This is not an easy question as there is no obvious way of producing f.

Open problem 25. What happens at the end points $\lambda = \frac{1}{s}$ and $\lambda = \frac{2}{s}$ in Theorems 7.11 and 7.12?

Open problem 26. What are the functions theoretic properties of the domains that arise as images of the unit disk under a global homeomorphism of finite distortion with integrability constraints on the distortion function? For the case of a quasidisk see [35].

7.5 Compactness

In this section we show that a certain class of mappings with exponentially integrable distortion is closed under weak convergence.

Theorem 7.15. *Let $n \geq 2$, $\lambda > 0$ and $A, B \geq 0$. Let $\mathscr{F}$ be the family of mappings $f : \Omega \to \mathbf{R}^n$ of finite distortion for which*

$$\int_\Omega J_f(x)\,dx \leq A \tag{7.6}$$

and

$$\int_\Omega e^{\lambda K_f(x)}\,dx \leq B . \tag{7.7}$$

Then for each $1 \leq p < n$ and every $f \in \mathscr{F}$ we have

(i) $\|Df\|^n_{L^p(\Omega)} \leq C_p(n, A, B) \int_\Omega J_f(x)\,dx$ and

(ii) $\mathscr{F}$ is closed under weak convergence in $W^{1,p}_{loc}(\Omega, \mathbf{R}^n)$.

Lemma 7.16. *The function*

$$F(x, y) = x^n y^{-1}$$

is convex on $(0, \infty) \times (0, \infty)$.

Proof. It is enough to show that

$$F(x, y) - F(a, b) \geq na^{n-1}b^{-1}(x - a) - a^n b^{-2}(y - b) \tag{7.8}$$

for every $x, y, a, b > 0$. From the arithmetic-geometric mean inequality we have

$$\left((x^n y^{-1})(a^n b^{-2} y)(a^n b^{-1})^{n-2}\right)^{\frac{1}{n}} \leq \frac{1}{n} x^n y^{-1} + \frac{1}{n} a^n b^{-2} y + \frac{n-2}{n} a^n b^{-1}$$

which can easily be rearranged to the desired inequality.   □

Let $\Phi_t(x) = t^{-n} \Phi(t^{-1} x)$, $t > 0$, be a standard approximation of unity; that is, $\Phi \in C_0^\infty(\mathbb{B})$, is non-negative and has integral 1. The convolution

$$\left(\mathscr{J}_f * \Phi_t\right)(a) = -\int_\Omega f_1(x) J\left(\Phi_t(a - \cdot), f_2, \ldots, f_n\right)(x) \, dx \qquad (7.9)$$

is a smooth function defined on the set $\Omega_t = \{a \in \Omega : \text{dist}(a, \partial\Omega) > t\}$. Now the following lemma provides us with a beneficial link between the distributional Jacobian $\mathscr{J}_f$ and the pointwise Jacobian J_f.

Lemma 7.17. *Let $f \in W^{1, \frac{n^2}{n+1}}(\Omega, \mathbf{R}^n)$. For almost every $a \in \Omega$ we have*

$$J_f(a) = \lim_{t \to 0} \left(\mathscr{J}_f * \Phi_t\right)(a). \qquad (7.10)$$

Proof. Let us recall that as in Remark 2.11 (a) we know by the Sobolev embedding Theorem A.18 that $f \in L_{\text{loc}}^{n^2}$ and hence $|f| \cdot |Df|^{n-1} \in L_{\text{loc}}^1$. Let us disclose in advance that the points $a \in \Omega$ for which we achieve (7.10) are determined by the properties

$$\lim_{t \to 0} \int_{B(a,t)} |Df(x) - Df(a)|^{\frac{n^2}{n+1}} \, dx = 0 \qquad (7.11)$$

and

$$\lim_{t \to 0} \frac{1}{t} \left(\int_{B(a,t)} |f(x) - f(a) - Df(a)(x - a)|^{n^2} \, dx\right)^{\frac{1}{n^2}} = 0. \qquad (7.12)$$

The first requirement is fulfilled at the Lebesgue points of $|Df|^{\frac{n^2}{n+1}}$. The second requirement is guaranteed at almost every point $a \in \Omega$ (see [29, Sect. 6.1.2]). We now split the integral at (7.9) as

$$\left(\mathscr{J}_f * \Phi_t\right)(a) = I_1 + I_2,$$

where

$$I_1 = -\int_\Omega \left[f_1(x) - f_1(a) - \langle \nabla f_1(a), x - a \rangle\right] J\left(\Phi_t(a - \cdot), f_2, \ldots, f_n\right)(x) \, dx$$

and

$$I_2 = -\int_\Omega \Big[f_1(a) + \langle \nabla f_1(a), x - a \rangle \Big] J\big(\Phi_t(a - \cdot), f_2, \ldots, f_n\big)(x)\, dx\,.$$

The first integral can be estimated by Hölder's inequality and using the fact that $|D\Phi_t(a - x)| \leq C(n)\, t^{-n-1}\, \chi_{B(a,t)}(x)$:

$$|I_1| \leq \frac{C(n)}{t} \Big(\fint_{B(a,t)} |f_1(x) - f_1(a) - \langle \nabla f_1(a), x - a \rangle|^{n^2}\, dx \Big)^{\frac{1}{n^2}}$$

$$\Big(\fint_{B(a,t)} |Df(x)|^{(n-1)\frac{n^2}{n^2-1}}\, dx \Big)^{\frac{n^2-1}{n^2}} \to 0 \tag{7.13}$$

by the requirements (7.12) and (7.11). Concerning the second term, we are allowed to integrate by parts (the integral is clearly finite and we can use standard limiting argument) to obtain

$$|I_2| = \int_\Omega \Phi_t(a - x) J\big(f_1(a) + \langle \nabla f_1(a), \cdot - a \rangle, f_2, \ldots, f_n\big)(x)\, dx$$

$$= J_f(a) + \int_\Omega \Phi_t(a - x)\Big[J\big(f_1(a) + \langle \nabla f_1(a), \cdot - a \rangle, f_2, \ldots, f_n\big)(x) - J_f(a) \Big]\, dx$$

where the latter integral converges to zero, as it is bounded by

$$C(n) \int_\Omega |\Phi_t(a - x)||Df_1(a)|\, |Df(x) - Df(a)|^{n-1}\, dx$$

$$\leq C(n)\, |Df_1(a)| \fint_{B(a,t)} |Df(x) - Df(a)|^{n-1}\, dx$$

$$\leq C(n)\, |Df_1(a)| \Big(\fint_{B(a,t)} |Df(x) - Df(a)|^{\frac{n^2}{n+1}}\, dx \Big)^{\frac{n^2-1}{n^2}} \to 0. \qquad \square$$

Proof (of Theorem 7.15). The uniform bound at (i) is rather simple. By Hölder's inequality and (7.7)

$$\|Df\|_{L^p(\Omega)}^n \leq \Big(\int_\Omega (KJ)^{\frac{p}{n}} \Big)^{\frac{n}{p}} \leq \|K\|_{L^{\frac{p}{n-p}}(\Omega)} \|J\|_{L^1(\Omega)} = C_p(n, B) \int_\Omega J_f(x)\, dx\,. \tag{7.14}$$

Let f_k be a sequence of mappings from our class. From (7.14) it is easy to see that f_k forms a bounded sequence in $W^{1,s}$ for each $s < n$. Hence we may assume passing to a subsequence that f_k converges to some f weakly in $W^{1,s}$ for every $s < n$ and thus for some subsequence strongly in L^q_{loc} for every $q < \infty$.

Since $Df_k \to Df$ weakly in $L^{\frac{n^2}{n+1}}$ and $f_k \to f$ in $L^{n^2}_{loc}$ we have that the distributional Jacobians $\mathscr{J}_{f_k}$ converge to $\mathscr{J}_f$ in $C_C^\infty(\Omega)$ i.e. for $\varphi \in C_0^\infty(\Omega)$ we have

$$\mathscr{J}_f(\varphi) = \lim_{k\to\infty} \mathscr{J}_{f_k}(\varphi) = \lim_{k\to\infty} \int_\Omega \varphi(x)\, J_{f_k}(x)\, dx, \qquad (7.15)$$

where in the last equality we use Theorem 2.12 to show that the pointwise Jacobian of $\{f_k\}$ equal to the distributional Jacobian. Therefore, if we take a nonnegative test function φ we get $\mathscr{J}_f(\varphi) \geq 0$. We can apply this to $\varphi(x) = \Phi_t(a - x)$ and by Lemma 7.17 we have

$$J_f(a) = \lim_{t\to 0} \mathscr{J}_f(\varphi) \geq 0$$

for almost every $a \in \Omega$. Therefore, f is an orientation preserving map i.e. $J_f(x) \geq 0$ almost everywhere in Ω.

Next we want to show that (7.15) remains valid for any bounded function $\varphi \in L^\infty(\Omega)$ with compact support. It is enough to consider test functions φ satisfying the bound

$$|\varphi(x)| \leq \chi_Q(x)$$

where χ_Q is the characteristic function of a cube Q and $2nQ \subset \Omega$. By Theorem 6.1 we know that for some $\alpha > 0$

$$\|J_{f_k}\|_{L^1 \log^\alpha L(2Q)} \leq M \qquad (7.16)$$

with M independent of k. We mollify φ by convolution with Φ_t, where t is chosen to be so small that $\varphi_t \in C_0^\infty(2Q)$. For given $T \geq 1$ we have

$$\int_{2Q} |\varphi_t(x) - \varphi(x)|\, J_{f_k}(x)\, dx \leq T\, \|\varphi_t - \varphi\|_{L^1(2Q)} + 2 \int_{\{x \in 2Q : J_{f_k}(x) \geq T\}} J_{f_k}(x)\, dx. \qquad (7.17)$$

In view of (7.16) we can choose T so big that the last integral is uniformly small. For this T we choose t so that $T\|\varphi_t - \varphi\|$ is small and combining this with (7.15) for φ_t we get

$$\lim_{k\to\infty} \int_\Omega J_{f_k}(x)\, \chi_Q(x)\, dx = \int_\Omega J_f(x)\, \chi_Q(x)\, dx \qquad (7.18)$$

as desired.

Next we will prove the critical lower semicontinuity property

$$\int_\Omega \eta(x)e^{\lambda K(x)}dx \le \liminf_{k\to\infty}\int_\Omega \eta(x)e^{\lambda K_{f_k}(x)}dx \qquad (7.19)$$

for each nonnegative test function $\eta \in L^\infty(\Omega)$ with compact support. Fix $\varepsilon > 0$ and write

$$K_f^\varepsilon(x) = \frac{|Df(x)|^n}{\varepsilon + J_f(x)}.$$

By Lemma 7.16 and (7.8) we obtain

$$\frac{|Df_k(x)|^n}{\varepsilon + J_{f_k}(x)} - \frac{|Df(x)|^n}{\varepsilon + J_f(x)} \ge n\frac{|Df(x)|^{n-1}}{\varepsilon + J_f(x)}\big(|Df_k(x)| - |Df(x)|\big)$$

$$-\frac{|Df(x)|^n}{(\varepsilon + J_f(x))^2}\big(J_{f_k}(x) - J_f(x)\big). \qquad (7.20)$$

Together with the convexity of the exponential function this implies

$$e^{\lambda K_{f_k}^\varepsilon(x)} - e^{\lambda K_f^\varepsilon(x)} \ge \lambda e^{\lambda K_f^\varepsilon(x)}\Big[n\frac{|Df(x)|^{n-1}}{\varepsilon + J_f(x)}\big(|Df_k(x)| - |Df(x)|\big)$$

$$-\frac{|Df(x)|^n}{(\varepsilon + J_f(x))^2}\big(J_{f_k}(x) - J_f(x)\big)\Big]. \qquad (7.21)$$

For $T > 0$ we set

$$E_T = \Big\{\Big|e^{\lambda K_f^\varepsilon(x)}\frac{|Df(x)|^n}{(\varepsilon + J_f(x))^2}\Big| \le T \text{ and } \Big|e^{\lambda K_f^\varepsilon(x)}\frac{|Df(x)|^{n-1}}{\varepsilon + J_f(x)}\Big| \le T\Big\}.$$

We now consider what happens in (7.21) when we first multiply it by non-negative $\eta \in L^\infty_{loc}(\Omega)$ and χ_{E_T}, then integrate over Ω, and let $k \to \infty$. The second term in the right-hand side converges to 0, because J_{f_k} converges to J_f weakly in $L^1_{loc}(\Omega)$, by (7.18). We fix the unit vector functions $\xi = \xi(x)$ and $\zeta = \zeta(x)$ such that

$$|Df(x)| = \langle Df(x)\xi(x), \zeta(x)\rangle.$$

Thus

$$|Df_k(x)| - |Df(x)| \ge \langle(Df_k(x) - Df(x))\xi(x), \zeta(x)\rangle. \qquad (7.22)$$

Using $\eta\xi\zeta \in L^\infty_{loc}$ and the assumption that $\{Df_k\}$ converges weakly in L^1 we obtain that the $\liminf_{k\to\infty}$ of the integral of the right-hand side is zero and thus the $\liminf_{k\to\infty}$ of the left-hand side integral is nonnegative. Altogether this implies

$$\int_\Omega \eta(x)\chi_{ET}(x)e^{\lambda K^\varepsilon(x)}dx \le \liminf_{k\to\infty}\int_\Omega \eta(x)\chi_{ET}(x)e^{\lambda K^\varepsilon_{f_k}(x)}dx$$

$$\le \liminf_{k\to\infty}\int_\Omega \eta(x)e^{\lambda K^\varepsilon_{f_k}(x)}dx. \tag{7.23}$$

Letting $\varepsilon \to 0$ and $T \to \infty$ we have (7.19). Together with (7.18) this implies that $\mathscr{F}$ is closed under weak convergence in $W^{1,p}$.

It remains to show that f has a finite distortion. The proof is very similar to that for inequality (7.19) and it is based on the inequality (7.20). Because of the uniform bound at (7.14) the sequence $\{f_k\}$ actually converges weakly in $W^{1,s}(\Omega, \mathbf{R}^n)$ for every $1 \le s < n$ and by the analysis as in (7.16)–(7.18) we have that J_{f_k} converges to J_f weakly in $L^1_{\text{loc}}(\Omega)$. Furthermore observing that Df_k converges weakly to Df in L^1_{loc}, we may pass to the limit in (7.20) multiplied by $\eta(x)\chi_{ET}(x)$ and we get

$$\int_\Omega \eta(x)\frac{|Df(x)|^n}{\varepsilon + J_f(x)}\chi_{ET}(x)\,dx \le \liminf_{k\to\infty}\int_\Omega \eta(x)\frac{|Df_k(x)|^n}{\varepsilon + J_{f_k}(x)}\chi_{ET}(x)\,dx$$

for every $\eta \in L^\infty_{\text{loc}}(\Omega)$. As before we let $\varepsilon \to 0$ and then $T \to \infty$. From the resulting inequality we may infer that

$$|Df(x)|^n \le M(x)\,J_f(x) \quad a.e.$$

for some measurable function $1 \le M(x) < \infty$ and this completes the proof of Theorem 7.15. $\square$

Remark 7.18. The results of this section were established by Iwaniec et al. in [63] where they have shown much more on this subject. For the significance of compactness arguments for the solvability of degenerate Beltrami equation see for example [5, 94] and [44].

Appendix

A.1 Simple Linear Algebra

For the vector $x \in \mathbf{R}^n$ we use $|x|$ to denote the Euclidean norm. For two vectors $x, y \in \mathbf{R}^n$ we denote their inner product by

$$\langle x, y \rangle := \sum_{i=1}^{n} x_i y_i .$$

For an $n \times n$-matrix A we use $|A|$ to denote the operator norm, i.e.

$$|A| = \sup\{|Ax| : \ x \in \mathbf{R}^n \text{ and } |x| \le 1\} .$$

It is easy to see that other norms are equivalent.

By I we denote the unit $n \times n$-matrix with 1 on the diagonal and 0 elsewhere. We use the notation adj A to denote the adjoint (or adjugate) matrix of matrix A. It contains the $(n-1) \times (n-1)$-subdeterminants of the matrix A and it satisfies the formula

$$A \operatorname{adj} A = I \det A, \tag{A.1}$$

where $\det A$ denotes the determinant of A.

A.2 Covering Theorems

We use covering theorems to select subcollections that consists of balls B_j that are disjoint or that have bounded overlap.

S. Hencl and P. Koskela, *Lectures on Mappings of Finite Distortion*, Lecture Notes in Mathematics 2096, DOI 10.1007/978-3-319-03173-6,
© Springer International Publishing Switzerland 2014

Theorem A.1 (Vitali). *Let $\mathscr{B}$ be a collection of closed balls in $\mathbf{R}^n$ such that*

$$\sup\{\operatorname{diam} B : B \in \mathscr{B}\} < \infty.$$

Then there are $B_1, B_2, \ldots$ (possibly a finite sequence) from this collection such that $B_i \cap B_j = \emptyset$ for $i \neq j$ and

$$\bigcup_{B \in \mathscr{B}} B \subset \bigcup_j 5B_j.$$

For a proof we refer the reader to [95]. Let us anyhow briefly explain the idea in a simple case. Suppose that the family $\mathscr{B}$ consists of balls $B(x, r_x)$, where $x \in A$ and A is bounded. Let $M = \sup_{x \in A} r_x$. Choose a ball $B_1 = B(x, r_x)$ so that $r_x > 3M/4$. Continue by considering points in $A \setminus 3B_1$, and repeating the first step (now letting $M_1 = \sup_{y \in A \setminus 3B_1} r_y$) and after that continue by induction.

In the Euclidean setting, a subcollection often can be chosen so that we only have uniformly bounded overlap for the cover.

Theorem A.2 (Besicovitch). *Let $\mathscr{B}$ be a collection of closed balls in $\mathbf{R}^n$ such that the set A consisting of the centers is bounded. Then there is a countable (possibly finite) subcollection $B_1, B_2, \ldots$ such that*

$$\chi_A(x) \leq \sum_j \chi_{B_j}(x) \leq C(n)$$

for all x.

In more general settings, say, in the Heisenberg group, Besicovitch fails. The reason it holds in the Euclidean setting, is basically the following geometric fact:

Suppose that we are given $B(x_1, r_1)$ and $B(x_2, r_2)$ so that $0 \in B(x_1, r_1) \cap B(x_2, r_2)$, $x_1 \notin B(x_2, r_2)$ and $x_2 \notin B(x_1, r_1)$. Then the angle between the vectors x_1 and x_2 is at least $60°$.

For a proof of the Besicovitch covering theorem, we again refer to [95].

A.3 L^p-Spaces

Recall that $L^p(\Omega)$, $1 \leq p < \infty$, consists of (equivalence classes) of measurable functions u with

$$\int_\Omega |u|^p < \infty.$$

We write

$$\|u\|_{L^p} = \|u\|_p := \left(\int_\Omega |u|^p \right)^{1/p}.$$

Furthermore, $L^\infty(\Omega)$ consists of those measurable functions on Ω that are essentially bounded. Then $\|u\|_{L^\infty} = \|u\|_\infty$ is the essential supremum of $|u|$ over Ω. If $1 < p < \infty$, we set $p' = p/(p-1)$, and we define $1' = \infty$. With this notation, we have the Minkowski

$$\|u + v\|_p \le \|u\|_p + \|v\|_p$$

and Hölder

$$\|uv\|_1 \le \|u\|_p \|v\|_{p'}$$

inequalities.

One often needs the following spherical coordinates. Given a Borel function $u \in L^1(B(0, 1))$ we have that

$$\int_{B(0,1)} u = \int_{S^{n-1}(0,1)} \int_0^1 u(tw) t^{n-1} \, dt dw.$$

Here we use the notation $\int_{S^{n-1}(c,t)} f(x) \, dx$ to denote integration with respect to the surface measure, which is a constant multiple of the Hausdorff measure $\mathscr{H}^{n-1}$.

We say that a sequence $\{u_i\}_i$ converges to u in $L^p(\Omega)$ if all these functions belong to $L^p(\Omega)$ and if $\|u - u_i\|_p \to 0$ when $i \to \infty$. We then write $u_i \to u$ in $L^p(\Omega)$. If $u_i \to u$ in $L^p(\Omega)$, then there is a subsequence $\{u_{i_k}\}_k$ of $\{u_i\}_i$ which converges to u pointwise almost everywhere. For $1 \le p < \infty$, continuous functions are dense in $L^p(\Omega)$: given $u \in L^p(\Omega)$ one can find continuous u_i with $u_i \to u$ both in $L^p(\Omega)$ and almost everywhere. This can be easily seen by first approximating u by simple functions, then approximating the associated measurable sets by compact sets and finally approximating the characteristic functions of the compact sets by continuous functions.

The dual of $L^p(\Omega)$ is $L^{p/(p-1)}(\Omega)$ when $1 < p < \infty$. Then

$$\|u\|_p = \sup_{\|\varphi\|_{\frac{p}{p-1}} = 1} \|u\varphi\|_1.$$

One of the inequalities easily follows by Hölder's inequality and the other by choosing φ to be a suitable constant multiple of $|u|^{p-1}$.

We also need the following weak compactness property: if $\{u_j\}_j$ is a bounded sequence in $L^p(\Omega)$, $1 < p < \infty$, then there is a subsequence $\{u_{j_k}\}_k$ and a function $u \in L^p(\Omega)$ so that

$$\lim_{k \to \infty} \int_\Omega u_{j_k} \varphi = \int_\Omega u\varphi$$

for each $\varphi \in L^{p/(p-1)}(\Omega)$. We then write

$$u_{j_k} \rightharpoonup u.$$

This notation should in principle include the exponent p, but the exponent in question is typically only indicated when its value is not obvious. This function u, called the weak limit, is unique and satisfies

$$\|u\|_p \leq \liminf_{k \to \infty} \|u_{j_k}\|_p.$$

The existence of the weak limit u follows from the fact that $L^p(\Omega)$, $1 < p < \infty$, is reflexive. Furthermore, the norm estimate on u is a consequence of a general result according to which a norm is lower semicontinuous with respect to the associated weak convergence. In general, weak convergence is defined by considering bounded linear mappings $T : X \to \mathbf{R}$; in the case of $L^p(\Omega)$, $1 < p < \infty$, they can be identified with elements of $L^{p/(p-1)}(\Omega)$. If $v_j = (v_1^j, \cdots, v_n^j) \in L^p(\Omega)$, then

$$v_j \rightharpoonup u$$

means that

$$v_i^j \rightharpoonup u_i$$

for each $1 \leq i \leq n$.

When we apply the above to a sequence $A_j(x)$ of $n \times n$-matrix functions, we conclude that the boundedness in $L^p(\Omega)$, $1 < p < \infty$ of the sequence $\{|A_j(x)|\}_j$ guarantees the existence of an $n \times n$-matrix function $A(x) \in L^p(\Omega)$ so that the rows (or columns) of a subsequence of $\{|A_j(x)|\}_j$ converge weakly to the corresponding rows (or columns) of $A(x)$. Notice that boundedness above is independent of the initial norm (like the operator or Hilbert-Schmidt one). Then $\|A\|_p \leq C_n \liminf_{k \to \infty} \|A_{j_k}\|_p$. In fact, one can show that

$$\|A\|_p \leq \liminf_{k \to \infty} \|A_{j_k}\|_p;$$

the L^p-norms generated by the operator or Hilbert-Schmidt norms are equivalent and so the associated concepts of weak convergence coincide.

We need the following sufficient condition for weak compactness in L^1.

Lemma A.3. *Let $\{g_j\}_{j \in \mathbf{N}}$ be a sequence of measurable functions on a domain $\Omega \subset \mathbf{R}^n$ of finite measure. Suppose that there is $H \in L^1(\Omega)$ such that for almost every $y \in \Omega$ and for every $j \in \mathbf{N}$ we have $|g_j(y)| \leq H(y)$. Then there is a subsequence $\{\tilde{g}_j\}_{j \in \mathbf{N}}$ of $\{g_j\}_{j \in \mathbf{N}}$ and $g \in L^1(\Omega)$ such that the subsequence $\{\tilde{g}_j\}_{j \in \mathbf{N}}$ converges weakly to g in $L^1(\Omega)$.*

Proof. We may assume that $H > 0$ everywhere on Ω. Define

$$h_j = \frac{g_j}{H}.$$

Since $0 \leq h_j \leq 1$, the sequence $\{h_j\}_{j \in \mathbb{N}}$ is bounded in $L^2(\Omega)$. Hence, after passing to a suitable subsequence, we may assume that $h_j \rightharpoonup h \in L^2(\Omega)$ in $L^2(\Omega)$. Thus

$$\int_\Omega h_j \eta \, dx \to \int_\Omega h \eta \, dx \tag{A.2}$$

for each $\eta \in L^2(\Omega)$. It is easy to show that $0 \leq h \leq 1$ a.e. and hence $h \in L^\infty$. Given $k \in \mathbb{N}$, set

$$H_k = \min\{H, k\}.$$

Let $\varphi \in L^\infty(\Omega)$. Then $H_k \varphi \in L^\infty(\Omega) \subset L^2(\Omega)$. By the triangle inequality

$$\left| \int_\Omega (g_j \varphi - hH\varphi) \, dx \right| = \left| \int_\Omega (h_j H\varphi - hH\varphi) \, dx \right|$$

$$\leq \left| \int_\Omega (h_j H_k \varphi - hH_k \varphi) \, dx \right|$$

$$+ \left| \int_\Omega (h_j \varphi - h\varphi)(H - H_k) \, dx \right|.$$

With the help of (A.2) we now obtain

$$\limsup_{j \to \infty} \left| \int_\Omega (g_j \varphi - hH\varphi) \, dx \right| \leq \limsup_{j \to \infty} \left| \int_\Omega (h_j \varphi - h\varphi)(H - H_k) \, dx \right|$$

$$\leq C \int_\Omega |H - H_k| \, dx.$$

Since $H \in L^1(\Omega)$, we conclude that

$$\lim_{j \to \infty} \int_\Omega g_j \varphi \, dx = \int_\Omega hH\varphi.$$

Since $hH \in L^1(\Omega)$, the claim follows. $\qquad\square$

On several occasions we will need Jensen's inequality.

Theorem A.4 (Jensen's Inequality). *Let $G \subset \mathbf{R}^n$ be a measurable set with $0 < |G| < \infty$ and let $\Phi : [0, \infty) \to [0, \infty)$ be a convex function. Then for every nonnegative measurable function $h : G \to [0, \infty)$ we have*

$$\Phi\left(\fint_G h(x)\,dx\right) \le \fint_G \Phi(h(x))\,dx\,.$$

A.4   Maximal Operator

Let $u \in L^1_{\text{loc}}(\mathbf{R}^n)$. The non-centered maximal function of u is

$$Mu(x) = \sup_{x \in B(y,r)} \fint_{B(y,r)} |u|.$$

Recall that for a measurable set A with $0 < |A| < \infty$,

$$\fint_A v = v_A = \frac{1}{|A|}\int_A v\,.$$

Remark A.5. (1) According to the Lebesgue differentiation theorem

$$Mu(x) \ge |u(x)|$$

almost everywhere.

(2) There are many other maximal functions. For example the restricted, centered maximal function

$$M^C_\delta u(x) = \sup_{0 < r < \delta} \fint_{B(x,r)} |u|.$$

(3) We always have $M^C_\infty u(x) \le Mu(x) \le 2^n M^C_\infty u(x)$.

(4) Notice that $\{Mu > t\}$ is open for each $t \ge 0$ and, consequently, Mu is measurable. Indeed, if $x \in \{Mu > t\}$, then it immediately follows from the definition that $B(y,r) \subset \{Mu > t\}$, for some $B(y,r)$ containing x.

Theorem A.6.   *If $u \in L^1(\mathbf{R}^n)$ and $t > 0$, then*

$$|\{Mu > t\}| \le \frac{5^n}{t}\int_{\{Mu>t\}} |u| \le \frac{5^n}{t}\|u\|_1. \tag{A.3}$$

Proof. We may assume that $M := \int_{\{Mu>t\}} |u| < \infty$. For each $x \in \{Mu > t\}$ there is a ball B such that $x \in B$ and

$$\fint_B |u| > t$$

and hence

$$|B| < t^{-1} \int_B |u| \, .$$

If $y \in B$, then $Mu(y) > t$ and thus $B \subset \{Mu > t\}$. So

$$|B| < \frac{1}{t} \int_B |u| \le \frac{1}{t} \int_{\{Mu>t\}\cap B} |u|.$$

By the Vitali covering theorem, Theorem A.1, we find pairwise disjoint balls $B_1, B_2, \ldots$ as above so that $\{Mu > t\} \subset \bigcup 5B_j$. Then

$$|\{Mu > t\}| \le \sum |5B_j| = 5^n \sum |B_j| \le \frac{5^n}{t} \sum \int_{B_j} |u| \le \frac{5^n}{t} \int_{\{Mu>t\}} |u|.$$

$\square$

The following lemma implies (by sending $\lambda \to 0$) that the maximal operator is bounded from L^p to L^p for $p > 1$.

Lemma A.7. *Let* $p > 1$, $\lambda > 0$ *and* $v \in L^p$. *Then*

$$\int_{\{Mv>\lambda\}} (Mv)^p \le C(n, p) \int_{\{|v|>\frac{\lambda}{2}\}} |v|^p.$$

Proof. Using the Fubini theorem and the estimate (A.3) we see that

$$\int_{\{Mv>\lambda\}} (Mv)^p = \int_{\{Mv>\lambda\}} \int_0^{Mv(x)} p t^{p-1} \, dt \, dx$$

$$= p \int_\lambda^\infty t^{p-1} |\{Mv > t\}| \, dt + \lambda^p |\{Mv > \lambda\}|$$

$$\le C \int_\lambda^\infty t^{p-2} \int_{\{|v|>\frac{t}{2}\}} |v(x)| \, dx \, dt + C \lambda^{p-1} \int_{\{|v|>\frac{\lambda}{2}\}} |v(x)| \, dx$$

$$\le C \int_\lambda^\infty \int_{\{|v|>\frac{t}{2}\}} |v(x)|^{p-1} \, dx \, dt + C \int_{\{|v|>\frac{\lambda}{2}\}} |v(x)|^p \, dx$$

$$= C \int_{\{|v|>\frac{\lambda}{2}\}} |v(x)|^{p-1} \int_\lambda^{2|v(x)|} dt \, dx + C \int_{\{|v|>\frac{\lambda}{2}\}} |v(x)|^p \, dx$$

$$\le C \int_{\{|v|>\frac{\lambda}{2}\}} |v(x)|^p \, dx.$$

$\square$

Remark A.8. Suppose that $u \in L^p(\Omega)$, $p > 1$. Applying the previous lemma to the zero extension of u we conclude that $\int_\Omega (Mu)^p \le C(p, n) \int_\Omega |u|^p$. Similarly, the inequality (A.3) can be restricted to Ω when $u \in L^1(\Omega)$.

The case $p = 1$ was not left out by accident from the previous lemma.

Example A.9. If $u(x) = \chi_{B(0,1)}(x)$, then Mu behaves like $\frac{C}{|x|^n}$ close to ∞ and hence $Mu \notin L^1(\mathbf{R}^n)$. In fact, $Mu \notin L^1(\mathbf{R}^n)$ unless u is the zero function.

We continue with a powerful tool from harmonic analysis, the Calderón-Zygmund decomposition, and some consequences of this decomposition.

The dyadic decomposition of a cube Q_0 consists of open cubes $Q \subset Q_0$ with faces parallel to the faces of Q_0 and of edge length $l(Q) = 2^{-i}l(Q_0)$, where $i = 1, 2, \ldots$ refer to the generation in the construction. The cubes in each generation cover Q_0 up to a set of measure zero and the closures of the cubes in a fixed generation cover Q_0; there are 2^{in} cubes of edge length $2^{-i}l(Q_0)$ in the ith generation and the cubes corresponding to the same generation are pairwise disjoint. For almost every $x \in Q_0$, there is a (unique) decreasing sequence $Q_0 \supset Q_1 \supset \cdots$ of cubes in the dyadic decomposition so that $\{x\} = \bigcap Q_i$. In what follows, Q, Q_0, Q_x etc. are cubes.

Theorem A.10 (Calderón-Zygmund Decomposition). *Let $Q_0 \subset \mathbf{R}^n$, $u \in L^1(Q_0)$, and suppose that*

$$0 \le \fint_{Q_0} u \le t.$$

Then there is a subcollection $\{Q_j\}$ from the dyadic decomposition of Q_0 so that $Q_i \cap Q_j = \emptyset$ when $i \ne j$,

$$t < \fint_{Q_j} u \le 2^n t$$

for each j, and $u(x) \le t$ for almost every $x \in Q_0 \setminus \bigcup Q_j$.

Proof. For almost every $x \in Q_0$ there is a decreasing sequence $\{Q_j\}$ of dyadic cubes so that $\{x\} = \bigcap Q_j$. By the Lebesgue differentiation theorem

$$\lim_{j \to \infty} \fint_{Q_j} u = u(x)$$

for almost every such x. Let $u(x) > t$ and assume that the above holds for x with the sequence $\{Q_j\}$. Then there must be maximal $Q_x := Q_{j(x)}$ so that

$$\fint_{Q_x} u > t.$$

For this cube we have

$$t < \fint_{Q_x} u \le 2^n \fint_{Q_{j(x)-1}} u \le 2^n t.$$

We can pick such a cube Q_x for almost every x with $u(x) > t$. It is then easy to choose the desired subcollection from the cubes Q_x.   □

The dyadic maximal function of a measurable function u (with respect to a cube Q_0) is defined by

$$M_{Q_0}u(x) = \sup_{x\in\overline{Q}\subset Q_0} \fint_Q |u|,$$

where the supremum is taken over all cubes Q that belong to the dyadic decomposition of Q_0 and whose closures contain x.

Remark A.11. As for the usual maximal function, we have the weak-type estimate

$$|\{x \in Q_0 :\ M_{Q_0}u(x) > t\}| \le \frac{2\cdot 5^n}{t} \int_{\{x\in Q_0 : |u(x)| > \frac{t}{2}\}} |u|$$

for the dyadic maximal function. Moreover,

$$\int_{Q_0} (M_{Q_0}u)^p \le C(p,n) \int_{Q_0} |u|^p$$

for $p > 1$. The proof of the weak type estimate is actually easier than for the usual maximal operator because no covering theorem is needed.

The following simple consequence of the Calderón-Zygmund decomposition is essentially the converse of the weak type estimate for the dyadic maximal function.

Lemma A.12. *Let $u \in L^1(Q_0)$ and suppose $t \ge \fint_{Q_0} |u|$. Then*

$$\int_{\{x\in Q_0 : |u(x)| > t\}} |u| \le 2^n t |\{x \in Q_0 : M_{Q_0}u(x) > t\}|.$$

Proof. By the Calderón-Zygmund decomposition we find pairwise disjoint cubes $Q_1, Q_2, \ldots$ so that

$$t < \fint_{Q_j} |u| \le 2^n t$$

for all j, and $|u(x)| \le t$ almost everywhere in $Q_0 \setminus \bigcup Q_j$. Then

$$\int_{\{x\in Q_0 : |u(x)| > t\}} |u| \le \sum \int_{Q_j} |u|$$

$$\le \sum 2^n t |Q_j|$$

$$\le 2^n t |\{x \in Q_0 : M_{Q_0}u(x) > t\}|,$$

because

$$M_{Q_0}u(x) \geq \int_{Q_j} |u| > t$$

for each $x \in Q_j$.   □

A.5 Sobolev Spaces

Definition A.13. Let $\Omega \subset \mathbf{R}^n$ be open and $u \in L^1_{\text{loc}}(\Omega)$. A function $v \in L^1_{\text{loc}}(\Omega, \mathbf{R}^n)$ is called a weak derivative of u if

$$\int_\Omega \varphi(x)v(x)dx = -\int_\Omega u(x)\nabla\varphi(x)dx$$

for every $\varphi \in C^\infty_C(\Omega)$. We refer to v by Du. For $1 \leq p \leq \infty$ we define the Sobolev space

$$W^{1,p}(\Omega) = \{u \in L^p(\Omega) : Du \in L^p(\Omega, \mathbf{R}^n)\}$$

and we define the norm

$$\|u\|_{W^{1,p}(\Omega)} = \left(\int_\Omega |u|^p + \int_\Omega |Du|^p\right)^{\frac{1}{p}}.$$

Further $W^{1,p}(\Omega, \mathbf{R}^n)$ refers to mappings $f : \Omega \to \mathbf{R}^n$ whose each component function f_j, $j = 1, \cdots, n$, belongs to $W^{1,p}(\Omega)$. The definitions of $W^{1,p}_{\text{loc}}(\Omega)$ and $W^{1,p}_{\text{loc}}(\Omega, \mathbf{R}^n)$ should then be obvious.

A function $\hat{u}$ is called a representative of $u \in W^{1,p}(\Omega)$ if $u = \hat{u}$ almost everywhere.

Definition A.14. Let $\Omega \subset \mathbf{R}^n$ be open and let $1 \leq p \leq \infty$. We define the Sobolev space of functions with zero boundary value $W^{1,p}_0(\Omega)$ as the collection of all functions $u \in W^{1,p}(\Omega)$ for which there are $u_j \in C^\infty_C(\Omega)$ such that $u_j \to u$ in $W^{1,p}(\Omega)$.

Theorem A.15 (Definitions of Sobolev Spaces). *Let $u \in L^p(\Omega)$, $1 \leq p < \infty$, $\Omega \subset \mathbf{R}^n$. Then the following are equivalent:*

(1) (ACL) The function u has a representative $\tilde{u}$ that is absolutely continuous on almost all line segments in Ω parallel to the coordinate axes and whose (classical) partial derivatives belong to $L^p(\Omega)$.

(2) (H) There is a sequence $\{\varphi_j\}_j \subset C^\infty(\Omega)$ so that $\varphi_j \to u$ in $L^p(\Omega)$ and $\{\nabla\varphi_j\}_j$ is Cauchy in $L^p(\Omega)$.

(3) **(W)** *The function u belongs to* $W^{1,p}(\Omega)$.

Proof (sketch).

(2) $\Rightarrow$ (1):   Passing to a subsequence, we may assume that $\{\varphi_j(x)\}_j$ converges for almost every x. We define

$$\tilde{u}(x) = \lim_{j \to \infty} \varphi_j(x)$$

whenever the limit exists, and set, say, $\tilde{u}(x) = 0$ for the remaining $x \in \Omega$. Then $\tilde{u}(x) = u(x)$ almost everywhere in Ω. Since $\nabla\varphi_j$ is Cauchy in L^p it is easy to see that $\nabla\varphi_j$ converge to Du in L^p.
We fix $[a_1, b_1] \times \ldots \times [a_n, b_n] \subset \Omega$. By the Fubini theorem we obtain that for $\mathscr{L}_{n-1}$-a.e. $[x_2, \ldots, x_n] \in [a_2, b_2] \times \ldots \times [a_n, b_n]$ we have

$$\int_I |Du - \nabla\varphi_j|^p \xrightarrow[j \to \infty]{} 0$$

where $I = [a_1, a_2] \times [x_2, \ldots, x_n]$ and moreover, we may assume that $\varphi_j(x) \to u(x)$ for $\mathscr{H}^1$-a.e. $x \in I$. Let us fix disjoint intervals $[c_i, d_i] \subset [a_1, a_2]$ such that $\varphi_j(c_i, x_2, \ldots, x_n) \to u(c_i, x_2, \ldots, x_n)$ and $\varphi_j(d_i, x_2, \ldots, x_n) \to u(d_i, x_2, \ldots, x_n)$. By the fundamental theorem of calculus applied to the functions φ_j we obtain

$$\left| \varphi_j(c_i, x_2, \ldots, x_n) - \varphi_j(d_i, x_2, \ldots, x_n) \right| \leq \int_{I_i} |\nabla\varphi_j|$$

where $I_i = [c_i, d_i] \times [x_2, \ldots, x_n]$. We let $j \to \infty$ and then sum over i to obtain

$$\sum_i \left| u(c_i, x_2, \ldots, x_n) - u(d_i, x_2, \ldots, x_n) \right| \leq \sum_i \int_{I_i} |Du| .$$

As $Du \in L^1$ this implies that u is absolutely continuous on I by the absolute continuity of the integral. More precisely, u has a representative which is absolutely continuous on I.

(1) $\Rightarrow$ (3):   Integration by parts is valid in one dimension for absolutely continuous functions. Hence we can integrate by parts over line segments and then use the Fubini theorem. The weak derivatives v_j are the classical partial derivatives of absolutely continuous functions.

(3) $\Rightarrow$ (2):   We use the (smooth) convolution approximation: Let

$$\psi_1(x) = \begin{cases} 0, & |x| \geq 1 \\ C \exp\left(\frac{1}{|x|^2-1}\right), & |x| < 1, \end{cases}$$

where C is chosen so that $\int_{\mathbf{R}^n} \psi_1\, dx = 1$. Define

$$\psi_\varepsilon(x) = \frac{1}{\varepsilon^n}\, \psi_1\left(\frac{x}{\varepsilon}\right). \qquad (A.4)$$

If $v \in L^p_{loc}$, set

$$v^\varepsilon(x) = (\psi_\varepsilon * v)(x) = \int_\Omega \psi_\varepsilon(x-y)v(y)\, dy,$$

when $B(x,\varepsilon) \subset\subset \Omega$. If $v \in L^p(\mathbf{R}^n)$, then $v^\varepsilon \to v$ in $L^p(\mathbf{R}^n)$. Indeed, this is easy to see for a continuous function v and for a general function we find a sequence of continuous functions that converge to it in L^p. Also $v^\varepsilon(x) \to v(x)$ when x is a Lebesgue point of u.

Fix $x \in \Omega$ and $\varepsilon > 0$ small compared to $\mathrm{dist}(x, \partial\Omega)$. Now

$$\frac{u^\varepsilon(x+he_i) - u^\varepsilon(x)}{h} = \frac{1}{\varepsilon^n}\int_\Omega \underbrace{\frac{1}{h}\left[\psi_1\left(\frac{x+he_i-y}{\varepsilon}\right) - \psi_1\left(\frac{x-y}{\varepsilon}\right)\right]}_{\xrightarrow[h\to 0]{}\ \frac{1}{\varepsilon}\frac{\partial\psi_1}{\partial x_i}\left(\frac{x-y}{\varepsilon}\right) = \varepsilon^n\frac{\partial\psi_\varepsilon}{\partial x_i}(x-y)} u(y)\, dy$$

$$\xrightarrow[h\to 0]{} \int_\Omega \frac{\partial\psi_\varepsilon}{\partial x_i}(x-y)\, u(y)\, dy$$

by the dominated convergence theorem:

$$\int_\Omega \left|\frac{1}{h}\left[\psi_1\left(\frac{x+he_i-y}{\varepsilon}\right) - \psi_1\left(\frac{x-y}{\varepsilon}\right)\right]u(y)\right|\, dy \le \frac{1}{\varepsilon}\int_\Omega \|\nabla\psi_1\|_{L^\infty}|u(y)|\, dy.$$

Thus there exists a derivative of u^ε and

$$\frac{\partial u^\varepsilon}{\partial x_i}(x) = \int_\Omega \frac{\partial\psi_\varepsilon}{\partial x_i}(x-y)\, u(y)\, dy$$

and because ψ_ε is smooth, we see that u^ε is C^1. By repeating this argument one can check that u_ε is actually C^∞. Moreover, when $u \in W^{1,p}$,

$$\frac{\partial u^\varepsilon}{\partial x_i}(x) = \int \frac{\partial\psi_\varepsilon(x-y)}{\partial x_i} u(y)\, dy$$

$$= -\int \frac{\partial\psi_\varepsilon(x-y)}{\partial y_i} u(y)\, dy$$

$$= \int \psi_\varepsilon(x-y)\, v_i(y)\, dy.$$

If $v_i \in L^p(\mathbf{R}^n)$, then this convolution sequence converges to v_i in $L^p(\mathbf{R}^n)$. When u is given on Ω, use a partition of unity to reduce the setting to that of $\mathbf{R}^n$. □

The weak derivative coincides with the usual derivative if both derivatives exist.

Corollary A.16. *Let $\Omega \subset \mathbf{R}^n$, $d \in \mathbf{N}$ and let $f \in W^{1,1}_{\mathrm{loc}}(\Omega, \mathbf{R}^d)$ be differentiable a.e. Then the classical derivative $\nabla f(x)$ equals to the weak derivative $Df(x)$ for a.e. $x \in \Omega$.*

Proof. From the proof of previous theorem $(1) \Rightarrow (3)$ we see that $Df(x)$ equals almost everywhere to the classical derivatives of the absolutely continuous representative. If $\nabla f(x)$ exists, then also classical partial derivatives exist at this point and they must equal a.e. to the derivatives of the absolutely continuous representative.

□

It is well known that a Sobolev function satisfies the Poincaré inequality (see [29, Sect. 5.6.1]).

Theorem A.17. *Let $B \subset \mathbf{R}^n$ be a ball. If $u \in W^{1,1}(B)$, then*

$$\int_B |u(x) - u_B|\, dx \leq C \operatorname{diam}(B) \int_B |Du(x)|\, dx.$$

Moreover, we also use the Sobolev-Poincaré inequality (see [29, Sect. 4.5.2]).

Theorem A.18. *Let $B \subset \mathbf{R}^n$ be a ball, $1 \leq p < n$ and $p^* = \frac{np}{n-p}$. If $u \in W^{1,p}(B)$, then $u \in L^{p^*}(B)$ and*

$$\left(\fint_B |u(x) - u_B|^{p^*}\, dx \right)^{\frac{1}{p^*}} \leq C \operatorname{diam}(B) \left(\fint_B |Du(x)|^p\, dx \right)^{\frac{1}{p}}.$$

The following theorem tells us that Sobolev functions are Hölder continuous for $p > n$ (see [122, proof of Theorem 2.4.4]).

Theorem A.19. *Let $u \in W^{1,p}(5B)$ and let $p > n$. Then*

$$|u(x) - u(y)| \leq C(n,p)|x - y|^{1-n/p} \left(\int_{B(x,2|x-y|)} |Du|^p \right)^{1/p}$$

for all Lebesgue points $x, y \in B$ of u.

Actually, with some work one can relax the assumption to $u \in W^{1,p}(B)$ and replace $B(x, 2|x - y|)$ with $B(x, 2|x - y|) \cap B$.

Regarding the Poincaré inequality, each function that satisfies a Poincaré inequality is in fact a Sobolev function. The proof of the following theorem is from [32].

Theorem A.20. *Let $\Omega \subset \mathbf{R}^n$ be open and let $u, g \in L^1_{\mathrm{loc}}(\Omega)$. Assume that there is $C > 0$ such that for every ball $B \subset \Omega$ we have*

$$\int_B |u(x) - u_B|\, dx \le C \operatorname{diam}(B) \int_B |g(x)|\, dx.$$

Then $u \in W^{1,1}_{\text{loc}}(\Omega)$.

Proof. Let $A \subset\subset \Omega$ be a fixed domain. First we construct approximations of u. Let $k \in \mathbf{N}$ be such that $\{x : \operatorname{dist}(x, A) < \frac{4n}{k}\} \subset \Omega$ and denote the $1/k$-grid in $\mathbf{R}^n$ by

$$G_k = \{z \in (\tfrac{1}{k}\mathbf{Z} \times \ldots \times \tfrac{1}{k}\mathbf{Z}) : \operatorname{dist}(z, A) < \tfrac{n}{k}\}\,.$$

Pick a partition of unity $\{\phi_z\}_{z \in G_k}$ such that

each $\phi_z : \mathbf{R}^n \to \mathbf{R}$ is continuously differentiable;

$\operatorname{spt} \phi_z \subset B(z, \frac{n}{k})$ and $|\nabla \phi_z| \le Ck$; $\qquad\qquad\qquad$ (A.5)

$$\sum_{z \in G_k} \phi_z(y) = 1 \text{ for every } y \in A.$$

Now we set

$$u_k(y) = \sum_{z \in G_k} \phi_z(y) u_{B(z, \frac{1}{k})} \text{ for every } y \in A. \qquad (A.6)$$

The supports of ϕ_z have bounded overlap and hence this sum is locally finite and $u_k \in C^1(A)$. It is not difficult to show that $u_k \to u$ in $L^1(A)$. Indeed, this is simple for a continuous function u by uniform continuity, and the general case follows by approximation by continuous functions.

Next we need to estimate the derivative of u_k. For $y \in A$ we choose $z_0 \in G_k$ so that $y \in B(z_0, \frac{n}{k})$. Since $\sum \phi_z = 1$ and we have a locally finite sum we may write

$$Du_k(y) = D\Big(\sum_{z \in G_\varepsilon} \phi_z(y)\big(u_{B(z, \frac{1}{k})} - u_{B(z_0, \frac{2n}{k})}\big)\Big) = \sum_{z \in G_k} D\phi_z(y)\big(u_{B(z, \frac{1}{k})} - u_{B(z_0, \frac{2n}{k})}\big)\,.$$
$$\qquad\qquad\qquad\qquad\qquad\qquad\qquad\qquad\qquad\qquad\qquad (A.7)$$

Since $y \in B(z_0, \frac{n}{k})$ it is easy to see that

$$\phi_z(y) \ne 0 \Rightarrow B(z, \frac{1}{k}) \subset B(z_0, \frac{2n}{k})\,.$$

Hence we may use (A.7) and (A.5) to estimate

$$|Du_k(y)| \leq \sum_{\{z \in G_k:\, \phi_z(y) \neq 0\}} Ck \left| u_{B(z,\frac{1}{k})} - u_{B(z_0,\frac{2n}{k})} \right|$$

$$\leq \sum_{\{z \in G_k:\, \phi_z(y) \neq 0\}} Ck \left| \fint_{B(z,\frac{1}{k})} (u(x) - u_{B(z_0,\frac{2n}{k})})\, dx \right|$$

$$\leq \sum_{\{z \in G_k:\, \phi_z(y) \neq 0\}} Ck \frac{1}{|B(z_0,\frac{2n}{k})|} \frac{|B(z_0,\frac{2n}{k})|}{|B(z,\frac{1}{k})|} \int_{B(z_0,\frac{2n}{k})} \left| u(x) - u_{B(z_0,\frac{2n}{k})} \right| dx\,.$$

Only a bounded number of terms above are nonzero and hence we can use our assumption and $B(z_0, \frac{2n}{k}) \subset B(y, \frac{3n}{k})$ to obtain

$$|Du_k(y)| \leq Ck \fint_{B(z_0,\frac{2n}{k})} \left| u(x) - u_{B(z_0,\frac{2n}{k})} \right| dx$$

$$\leq C \fint_{B(z_0,\frac{2n}{k})} |g(x)|\, dx \leq C \fint_{B(y,\frac{3n}{k})} |g(x)|\, dx\,. \tag{A.8}$$

Since $\fint_{B(y,\frac{3n}{k})} |g| \to |g(y)|$ in $L^1(A)$ as $k \to \infty$, there is a subsequence $k_j \to 0$ such that $\fint_{B(y,\frac{3n}{k_j})} |g|$ has a majorant $H \in L^1(A)$. From this, (A.8) and Lemma A.3, we obtain that there is a subsequence $k_i \to \infty$ and $g \in L^1(A, \mathbf{R}^n)$ such that $Du_{k_i} \to g$ weakly in L^1. Since $u_k \in C^1$ we have

$$\int_A Du_{k_i}(y)\varphi(y)dy = -\int_A u_{k_i}(y)D\varphi(y)dy \tag{A.9}$$

for every test function $\varphi \in C_c^\infty(A)$. Since $u_k \to u$ in L^1, we obtain, after passing to the limit, that

$$\int_A g(y)\varphi(y)dy = -\int_A u(y)D\varphi(y)dy$$

which means that g is a weak gradient of u in A and therefore $u \in W^{1,1}(A)$. □

There is also a version of the previous theorem for BV-functions (recall that BV-functions were defined in Definition 5.1).

Theorem A.21. *Let $\Omega \subset \mathbf{R}^n$ be open, $u \in L^1_{\mathrm{loc}}(\Omega)$ and let μ be a Radon measure on Ω. Assume that there is $C > 0$ such that for every ball $B \subset \Omega$ we have*

$$\int_B |u(x) - u_B|\, dx \leq C\, \mathrm{diam}(B)\mu(B)\,.$$

Then $u \in BV_{\mathrm{loc}}(\Omega)$.

Proof. We proceed similarly to the proof of the previous theorem. Again we fix $A \subset\subset \Omega$ and we define u_k by the same formula (A.6). Again $u_k \in C^1$ and they converge in L^1 to u. Analogously to the estimate (A.8) we obtain using our assumption that

$$|Du_k(y)| \leq Ck \fint_{B(z_0, \frac{2n}{k})} \left|u(x) - u_{B(z_0, \frac{2n}{k})}\right| dx \leq C \frac{\mu(B(z_0, \frac{2n}{k}))}{|B(z_0, \frac{2n}{k})|}.$$

By the integration of this inequality over A we obtain

$$\int_A |Du_k(y)| \, dy \leq C \sum_{z \in G_k} \mu(B(z, \tfrac{2n}{k})) \leq C\mu(\{x \in \Omega : \text{dist}(x, A) < \tfrac{3n}{k}\}).$$

It follows that Du_k form a bounded sequence in L^1. Recalling that $u_k \to u$ in L^1, we conclude [29, paragraph 5.2.3] that $u \in BV(A)$. In fact, there is a subsequence and vector v of Radon measures such that Du_{k_i} converge to v weak star in measures. As before we have (A.9) and by passing to a limit we have

$$\int_A \varphi(y)dv(y) = -\int_A u(y)D\varphi(y)dy$$

which means that v is a weak gradient of u in A and therefore $u \in BV(A)$.   □

We have seen in Theorem A.15 that $W^{1,1}$-functions can be characterized using the ACL-property. Similarly, it is possible to characterize BV-functions using the BVL-property, i.e. that the function in question has bounded variation on almost all lines parallel to coordinate axes (see [2, Sect. 3.11]).

More precisely, let $i \in \{1, 2, \ldots, n\}$, $Q_0 = (0, 1)^n$ and by π_i denote the projection to the hyperplane perpendicular to i-th coordinate axis. For $y \in \pi_i(Q_0)$ we denote $g_{i,y}(t) = g(y + te_i)$.

Theorem A.22. *Let $g \in L^1(Q_0)$. Then $g \in BV(Q_0)$ if and only if for every $i \in \{1, \ldots, n\}$ the function $g_{i,y}(t) \in BV((0, 1))$ for $\mathcal{H}^{n-1}$-almost every $y \in \pi_i(Q_0)$ and moreover*

$$\int_{\pi_i(Q_0)} |Dg_{i,y}|((0, 1)) \, d\mathcal{H}^{n-1}(y) < \infty,$$

where $|Dg_{i,y}|((0, 1))$ denotes the total variation of our BV-function of a single variable. In this case we can estimate the total variation of Dg by

$$|Dg|(Q_0) \leq C \sum_{i=1}^n \int_{\pi_i(Q_0)} |Dg_{i,y}|((0, 1)) \, d\mathcal{H}^{n-1}(y).$$

A.6 Lipschitz Approximation of Sobolev Functions

We know by Theorem A.15 that, given a Sobolev function u, we can find C^1-smooth functions u_k such that $u_k \to u$ in $W^{1,1}$, but in some applications this approximation is not good enough. In this section we construct Lipschitz functions u_k that converge to u in $W^{1,1}$ and moreover $u = u_k$ on a big set. First we need a couple of lemmata.

Lemma A.23 (McShane Extension). *Let $A \subset \mathbf{R}^n$ and $f : A \to \mathbf{R}^m$ be L-Lipschitz, that is*

$$|f(x) - f(y)| \le L|x - y|$$

for all $x, y \in A$. Then there exists a $(\sqrt{m}L)$-Lipschitz $\tilde{f} : \mathbf{R}^n \to \mathbf{R}^m$ such that $\tilde{f}|_A = f$.

Proof. Let $m = 1$. Define

$$\tilde{f}(x) = \inf_{a \in A}\{f(a) + L|x - a|\}.$$

Then $\tilde{f}(x) = f(x)$ when $x \in A$: Since f is L-Lipschitz on A,

$$f(x) \le f(a) + L|x - a| \quad \text{when } x, a \in A,$$

and so $f(x) \le \tilde{f}(x)$. By the choice $a = x$ in the definition of $\tilde{f}(x)$ we obtain $\tilde{f}(x) \le f(x)$.

Given $x, y \in \mathbf{R}^n$, we have that

$$\tilde{f}(x) = \inf_{a \in A}\{f(a) + \underbrace{L|x - a|}_{\le L(|y-a|+|y-x|)} \}$$

$$\le L|y - x| + \tilde{f}(y).$$

Because this also holds with x replaced by y, we conclude that $\tilde{f}$ is L-Lipschitz.

Let us then consider the case $m \ge 2$. For given $f = (f_1, \ldots, f_m)$ define $\tilde{f} = (\tilde{f}_1, \ldots, \tilde{f}_m)$ as in the previous case. Now

$$|\tilde{f}(x) - \tilde{f}(y)|^2 = \sum_1^m |\tilde{f}_i(x) - \tilde{f}_i(y)|^2 \le mL^2|x - y|^2,$$

and the claim follows.   □

Remark A.24. By choosing a suitable extension different from the McShane extension, one could require above $\tilde{f}$ to be L-Lipschitz. This can be done using the so-called Kirszbaum extension.

In the following theorem we use the modified maximal function

$$M_{3r_0} u(x) = \sup_{x \in B(y,r) \subset B(x_0, 3r_0)} \fint_{B(y,r)} |u| \, .$$

Lemma A.25. *Let $B = B(x_0, r_0)$ be a ball in $\mathbf{R}^n$ and let $u \in W^{1,1}(B(x_0, 3r_0))$. For $\lambda > 0$ we define*

$$F_\lambda = \{x \in B : M_{3r_0} |Du(x)| < \lambda\} \cap \{x \in B : \ x \ is \ a \ Lebesgue \ point \ of \ u\}. \tag{A.10}$$

There is a constant $C > 0$ such that,

$$|u(x) - u(y)| \le C\lambda |x - y| \ for \ all \ x, y \in F_\lambda.$$

Moreover, the measure of the remaining set satisfies

$$|B \setminus F_\lambda| = o(\tfrac{1}{\lambda}) \, .$$

Proof. Let $x, y \in F_\lambda$. Choose $B_j = B(x, 2^{-j}|x - y|)$ for $j \ge 0$ and $B_j = B(y, 2^{j+1}|x-y|)$ for $j < 0$. As x and y are Lebesgue points we obtain $u_{B_j} \to u(x)$ as $j \to \infty$ and $u_{B_j} \to u(y)$ as $j \to -\infty$ and hence

$$|u(x) - u(y)| \le \sum_{-\infty}^{\infty} |u_{B_{j+1}} - u_{B_j}| \, .$$

Moreover, for $j > 0$ (and thus $B_{j+1} \subset B_j$) we can estimate the difference

$$|u_{B_{j+1}} - u_{B_j}| = \left| \frac{1}{|B_{j+1}|} \int_{B_{j+1}} (u(x) - u_{B_j}) \, dx \right|$$

$$\le \frac{1}{|B_{j+1}|} \int_{B_{j+1}} |u(x) - u_{B_j}| \, dx \le \frac{C(n)}{|B_j|} \int_{B_j} |u(x) - u_{B_j}| \, dx \tag{A.11}$$

and we have similar estimate for $j < 0$. For $|u_{B_0} - u_{B_1}|$ we add and subtract the term $u_{B_0 \cap B_1}$ and easily obtain the bound

$$\frac{1}{|B_0|} \int_{B_0} |u(x) - u_{B_0}| \, dx + \frac{1}{|B_1|} \int_{B_1} |u(x) - u_{B_1}| \, dx.$$

Hence we can use the Poincaré inequality, Theorem A.17, to obtain

$$|u(x) - u(y)| \leq \sum_{-\infty}^{\infty} |u_{B_{j+1}} - u_{B_j}| \leq \sum_{-\infty}^{\infty} C(n) \fint_{B_j} |u - u_{B_j}|$$

$$\leq C(n) \sum_{-\infty}^{\infty} r_j \fint_{B_j} |Du|$$

$$\leq C(n)|x - y|\big(\tilde{M}_{3r_0}|Du(x)| + \tilde{M}_{3r_0}|Du(y)|\big)$$

$$\leq 2C(n)|x - y|\lambda .$$

Thus we have $C(n)\lambda$-Lipschitz continuity on the set F_λ. By Remark A.8 and Theorem A.6 we have

$$|B \setminus F_\lambda| \leq \frac{5^n 2}{\lambda} \underbrace{\int_{\{M_{3r_0}|Du(z)|>\lambda\}\cap 3B} |Du|}_{\underset{\lambda\to\infty}{\longrightarrow} 0} = o\left(\tfrac{1}{\lambda}\right)$$

and the claim follows.   □

Remark A.26. The above proof shows that u is $C(n)\lambda$-Lipschitz in F_λ, where $|B \setminus F_\lambda| = o\left(\tfrac{1}{\lambda}\right)$. Use the McShane extension theorem to extend the restriction of u to this set as a $C(n)\lambda$-Lipschitz function u_λ, defined in entire B. Then

$$\int_B |Du - Du_\lambda| \leq \int_{B\setminus F_\lambda} (|Du| + |Du_\lambda|) \leq \int_{B\setminus F_\lambda} |Du| + C(n)\lambda o\left(\tfrac{1}{\lambda}\right) \underset{\lambda\to\infty}{\longrightarrow} 0$$

because

$$Du_\lambda(x) = Du(x) \tag{A.12}$$

at almost every point $x \in F_\lambda$.

Reason: If $E \subset \Omega$ is measurable, $\partial_i v$ and $\partial_i w$ exist almost everywhere in E and $v = w$ on E, then $\partial_i v = \partial_i w$ almost everywhere in E. Simply notice that almost every point x of E is of linear density one in the x_i-direction.

One can do even better. Consider the set

$$\mathrm{Bad}_\lambda' = \{x \in B : \tilde{M}_{3r_0} u(x) \geq \lambda\}.$$

Then $|\mathrm{Bad}_\lambda'| = o\left(\tfrac{1}{\lambda}\right)$. So, when λ is large, the distance from any point in Bad_λ' to $B \setminus \mathrm{Bad}_\lambda'$ is at most one. Thus the McShane extension u_λ of u from $F_\lambda \setminus \mathrm{Bad}_\lambda'$ is $C(n)\lambda$-Lipschitz and bounded in absolute value by $2C(n)\lambda$ on B. It follows that

$$\int_B |u - u_\lambda| + |Du - Du_\lambda| \underset{\lambda\to\infty}{\longrightarrow} 0.$$

The final estimate of the preceding remark yields the following corollary:

Corollary A.27. *If $u \in W^{1,1}(3B)$, then there is a sequence $\{u_j\}_{j=1}^{\infty}$ of Lipschitz functions such that*

$$\{x \in B : u_{j+1}(x) \neq u(x)\} \subset \{x \in B : u_j(x) \neq u(x)\}, \quad |\{x \in B : u_j(x) \neq u(x)\}| \overset{j \to \infty}{\to} 0$$

and

$$\int_B |u - u_j| + |Du - Du_j| \overset{j \to \infty}{\to} 0.$$

A.7 Differentiability and Approximative Differentiability

The following differentiability result due to Menchoff was independently also proved by Gehring and Lehto.

Lemma A.28. *Suppose that $f : \mathbf{R}^2 \to \mathbf{R}^2$ is a homeomorphism that belongs to $W_{\text{loc}}^{1,1}(\mathbf{R}^2, \mathbf{R}^2)$. Then f is differentiable almost everywhere.*

Proof. It suffices to prove the claim for the two component functions of f. So, let $u : \mathbf{R}^2 \to \mathbf{R}$ be one of them. Then, by the Lebesgue differentiation theorem,

$$\lim_{r \to 0} \frac{1}{|B(x_0, r)|} \int_{B(x_0, r)} |Du(x) - Du(x_0)| dx = 0 \tag{A.13}$$

for almost every x_0. Fix such a point x_0, let $r > 0$, and suppose $x \in B(x_0, r)$. Given an open rectangle R_x contained in $B(x_0, 2r)$ with $x \in R_x$, we define a function $v : \overline{R}_x \to \mathbf{R}$ by setting

$$v(y) = |u(y) - u(x_0) - \langle Du(x_0), x - x_0 \rangle|.$$

Since f is a homeomorphism, the maximum of v on $\overline{R}_x$ occurs on ∂R_x, say at $y_x^R \in \partial R_x$. We conclude that

$$|u(x) - u(x_0) - \langle Du(x_0), x - x_0 \rangle| \leq v(y_x^R)$$
$$\leq |u(y_x^R) - u(x_0) - \langle Du(x_0), y_x^R - x_0 \rangle|$$
$$+ |Du(x_0)||y_x^R - x|.$$

Thus, differentiability of u at x_0 follows if, given $\varepsilon > 0$, we can find $r > 0$ so that, for each $x \in B(x_0, r)$ we can make the right-hand side above smaller than εr, via a suitable choice of R_x.

Towards this end, we first deal with the term $|u(y_x^R) - u(x_0) - \langle Du(x_0), y_x^R - x_0 \rangle|$. We may assume that $Du(x_0) = 0$ by replacing u with the function w defined by setting $w(y) = u(y) - \langle Du(x_0), y - x_0 \rangle$. By yet other replacements, we may assume that $x_0 = 0 = u(x_0)$. Then (A.13) guarantees that

$$\lim_{r \to 0} \frac{1}{r^2} \int_{-r}^{r} \int_{-r}^{r} |Du(s,t)| \, ds \, dt = 0.$$

Set $A_r = \{-r < s < r : \int_{-r}^{r} |Du(s,t)| \, dt \geq \varepsilon r\}$ and $A^r = \{-r < t < r : \int_{-r}^{r} |Du(s,t)| \, ds \geq \varepsilon r\}$. Then, for each $\varepsilon > 0$, we can find $r_\varepsilon > 0$ so that

$$|A_r| \leq \varepsilon r \text{ and } |A^r| \leq \varepsilon r \tag{A.14}$$

for all $0 < r < r_\varepsilon$. By Theorem A.15 we know that u is absolutely continuous on almost all lines parallel to coordinate axes and hence we may assume that u is absolutely continuous on $y_1 \times [-r, r]$ for every $y_1 \notin A_r$ and on $[-r, r] \times y_2$ for every $y_2 \notin A^r$. We conclude that $|u(y_1, y_2)| \leq 2\varepsilon r$ provided $y_1 \notin A_r$, $y_2 \notin A^r$ and $0 < r < r_\varepsilon$. Thus we have found plenty of rectangles R_x, on the boundary of whose, $|u(y_x^R) - u(x_0) - \langle Du(x_0), y_x^R - x_0 \rangle| \leq 2\varepsilon r$.

By (A.14) we may moreover assume that the side-length of each rectangle R_x is at most εr. Finally,

$$|Du(x_0)| |y_x^R - x| \leq (\text{dist}(x, \partial R_x) + \text{diam}(R_x)) |Du(x_0)|$$

and hence this term is handled by the estimate for the size of R_x. $\qquad\square$

For the study of the regularity of the inverse we need the following elementary observation.

Lemma A.29. *Let $f : \Omega \to \mathbf{R}^n$ be a homeomorphism which is differentiable at $x \in \Omega$ with $J_f(x) > 0$. Then f^{-1} is differentiable at $f(x)$ and $Df^{-1}(f(x)) = (Df(x))^{-1}$.*

Proof. Let us denote $y = f(x)$. We know that

$$\lim_{h \to 0} \frac{f(x+h) - f(x) - Df(x)h}{\|h\|} = 0 \tag{A.15}$$

and we would like to show that

$$\lim_{t \to 0} \frac{f^{-1}(y+t) - f^{-1}(y) - (Df(x))^{-1}t}{\|t\|} = 0.$$

Given $t \in \mathbf{R}^n$ we set $h = f^{-1}(y+t) - f^{-1}(y)$ which implies that

$$f(x+h) - f(x) = f(x + f^{-1}(y+t) - x) - f(x) = t.$$

By (A.15) and $J_f(x) > 0$ we obtain for small enough $\|h\|$ that

$$\|t\| = \|f(x+h) - f(x)\| \approx \|Df(x)h\| \approx \|h\| \, .$$

Now (A.15) implies that

$$0 = \lim_{h \to 0} \frac{(f(x+h) - f(x) - Df(x)h)(Df(x))^{-1}}{\|h\|}$$

which implies that

$$0 = \lim_{t \to 0} \frac{(Df(x))^{-1}(f(x+h) - f(x)) - h}{\|t\|}$$

$$= \lim_{t \to 0} \frac{(Df(x))^{-1}t - f^{-1}(y+t) - f^{-1}(y)}{\|t\|}$$

and we get the desired conclusion. $\square$

Definition A.30. Let $\Omega \subset \mathbf{R}^n$, $d \in \mathbf{N}$ and let $f : \Omega \to \mathbf{R}^d$ be a mapping. We say that f is approximatively differentiable at $x \in \Omega$ with approximative derivative $Df(x)$ if there is a set $A \subset \Omega$ of density one at x, i.e. $\lim_{r \to 0} \frac{|A \cap B(x,r)|}{|B(x,r)|} = 1$, such that

$$\lim_{y \to x, y \in A} \frac{f(y) - f(x) - Df(x)(y - x)}{\|y - x\|} = 0 \, .$$

It is well known that Sobolev functions are differentiable a.e.

Theorem A.31. Let $\Omega \subset \mathbf{R}^n$, $d \in \mathbf{N}$ and let $f \in W^{1,1}(\Omega, \mathbf{R}^d)$. Then f is approximatively differentiable a.e. in Ω and its approximative derivative coincides with its weak derivative a.e.

Proof. By Corollary A.27 we know that we can find a sequence of Lipschitz mappings $f_j : \Omega \to \mathbf{R}^d$ such that for

$$A_j := \{x \in \Omega : f_j(x) \neq f(x)\} \text{ we have } A_{j+1} \subset A_j \text{ and } |A_j| \to 0 \, .$$

The Lipschitz mappings f_j are differentiable a.e. (see Theorem 2.23) and their classical derivatives coincide with the weak derivatives Df_j (see Corollary A.16).

It is easy to see that $\Omega = \bigcup_j A_j \cup S$ where $|S| = 0$. Almost every point of A_j is a point of density of A_j and hence for almost every $x \in \Omega$ we can find j such that

$$\lim_{r \to 0} \frac{|A_j \cap B(x,r)|}{|B(x,r)|} = 1 \, .$$

Let us pick a point $x \in \Omega$ such that x is a point of density of A_j and f_j is differentiable at x. Since $f = f_j$ on A_j we obtain from the differentiability of f_j that

$$
\begin{aligned}
0 &= \lim_{y \to x} \frac{f_j(y) - f_j(x) - Df_j(x)(y - x)}{\|y - x\|} \\
&= \lim_{y \to x, \ y \in A_j} \frac{f(y) - f(x) - Df_j(x)(y - x)}{\|y - x\|}
\end{aligned}
$$

which implies the approximative derivative of f at x exists and equals to $Df_j(x)$.

$\square$

We need an analogue of Lemma A.29 for approximatively differentiable mappings. For its proof we need the following observation.

Lemma A.32. *Let $\Omega \subset \mathbf{R}^n$ and let $g : \Omega \to \mathbf{R}^n$ be Lipschitz mapping that is differentiable at x. Suppose that $J_g(x) \neq 0$ and let A be a set of density 1 at $x \in \Omega$. Then the density of the set $g(A)$ at $g(x)$ is 1.*

Proof. Let us denote $y = g(x)$ and let $r > 0$ be small enough such that $B(y,r) \subset f(\Omega)$. Since g is differentiable at x, continuous and $J_g(x) \neq 0$ we obtain that for r small enough we have

$$
B(y,r) \subset g\big(B(x, C_1 r)\big) \quad \text{for} \quad C_1 = 2|(Dg(x))^{-1}| . \tag{A.16}
$$

We know that

$$
\lim_{\tilde{r} \to 0} \frac{|A \cap B(x, \tilde{r})|}{|B(x, \tilde{r})|} = 1
$$

and hence for given $\varepsilon > 0$ we obtain that for small enough r we have

$$
\big|B(x, C_1 r) \setminus A\big| < \varepsilon |B(x, C_1 r)| = \varepsilon C_1^n |B(x, r)| .
$$

Since g is Lipschitz with constant L, (A.16) gives

$$
\big|B(y,r) \setminus g(A)\big| \leq \big|g(B(x, C_1 r)) \setminus g(A)\big| \leq \big|g\big(B(x, C_1 r) \setminus A\big)\big| < L^n \varepsilon C_1^n |B(x, r)| .
$$

This shows that

$$
\lim_{r \to 0} \frac{|g(A) \cap B(y, r)|}{|B(y, r)|} = 1 . \qquad \square
$$

Now we prove the desired analog of Lemma A.29, following [33, Lemma 2.1].

Lemma A.33. *Let* $f : \Omega \to \Omega'$ *be a homeomorphism such that* $f \in W^{1,1}(\Omega, \Omega')$ *and* $f^{-1} \in W^{1,1}(\Omega', \Omega)$. *Set*

$$E := \{y \in \Omega' : \ f^{-1} \text{ is approximatively differentiable at } y \text{ and } |J_{f^{-1}}(y)| > 0\} .$$

Then, there exists a Borel set $A \subset E$ *with* $|E \setminus A| = 0$ *such that*

$$f^{-1}(A) \subset F := \{x \in \Omega : \ f \text{ is approximatively differentiable at } x \text{ and } |J_f(x)| > 0\}$$

and we have

$$Df^{-1}(y) = \left(Df(f^{-1}(y))\right)^{-1} \text{ for every } y \in A .$$

Proof. From the proof of Theorem A.31 we know that there is a null set N_E such that for every $y \in E \setminus N_E$ we can find a Lipschitz mapping h and set $\tilde{E} \subset E$ such that $\tilde{E}$ has density one at y, $f^{-1} = h$ on $\tilde{E}$, h is differentiable at y and $Df^{-1}(y) = Dh(y)$. We can also require that $E \setminus N_E$ is a Borel set. Analogously, we can find a null set N_F so that for every $x \in F \setminus N_F$ we can find a Lipschitz mapping g and set $\tilde{F} \subset F$ such that $\tilde{F}$ has density one at x, $f = g$ on $\tilde{F}$, g is differentiable at x and $Df(x) = Dg(x)$.

By Corollary A.36 we obtain that there is a Borel set $J \subset \{J_f = 0\}$ such that $|f(J)| = 0$ and $|J| = |\{J_f = 0\}|$. This and the a.e. approximative differentiability of f (see Theorem A.31) shows that the set $N_F \cup (\Omega \setminus (F \cup J))$ has measure zero. We can find a Borel set N of measure zero such that $N_F \cup (\Omega \setminus (F \cup J)) \subset N$ and by the Area formula (A.19) applied to f^{-1} we know that

$$\int_{f(N)} |J_{f^{-1}}(y)| \, dy \le \int_{f^{-1}(f(N))} dx = |N| = 0 .$$

Since $J_{f^{-1}} > 0$ on E, this implies that $|f(N) \cap E| = 0$. For the set

$$A := E \setminus (f(N) \cup N_E \cup f(J))$$

we thus have $|E \setminus A| = 0$ and $f^{-1}(A) \subset F$.

Given $y \in A$ we denote $x = f^{-1}(y) \in F$ and we can find functions h, g and sets $\tilde{E}$ and $\tilde{F}$ as in the first paragraph. Since g is differentiable at x and h is differentiable at y we obtain

$$D(g \circ h)(y) = Dg(h(y))Dh(y) = Dg(x)Dh(y) . \tag{A.17}$$

Since $|J_g(x)| = |J_f(x)| > 0$ we obtain by Lemma A.32 that the set $g(\tilde{F}) \cap \tilde{E}$ has density one at y. For every $z \in g(\tilde{F}) \cap \tilde{E}$ we have $g(h(z)) = f(f^{-1}(z)) = z$ and hence $D(g \circ h)(y) = I$. Now (A.17) implies $Dh(y) = (Dg(x))^{-1}$. $\qquad\square$

A.8 Area and Coarea Formula

Let us recall the definition of the Lusin (N) condition.

Definition A.34. Let $\Omega \subset \mathbf{R}^n$ be open. We say that $f : \Omega \to \mathbf{R}^n$ satisfies the Lusin (N) condition if

$$\text{for each } E \subset \Omega \text{ such that } |E| = 0 \text{ we have } |f(E)| = 0 \, .$$

Theorem A.35. Let $f \in W^{1,1}_{\mathrm{loc}}(\Omega, \mathbf{R}^n)$ and let η be a nonnegative Borel measurable function on $\mathbf{R}^n$. Then

$$\int_\Omega \eta(f(x)) \, |J_f(x)| \, dx \le \int_{\mathbf{R}^n} \eta(y) \, N(f, \Omega, y) \, dy \, , \qquad (A.18)$$

where the multiplicity function $N(f, \Omega, y)$ of f is defined as the number of preimages of y under f in Ω. Moreover, there is an equality in (A.18) if we assume in addition that f satisfies the Lusin (N) condition.

Proof (Sketch of the Proof). It is known that each $f \in W^{1,1}$ is a.e. approximatively differentiable (see Theorem A.31), and by Corollary A.27 we can decompose $\Omega = S \cup \bigcup_{n=1}^\infty \Omega_i$ in a such a way that $|S| = 0$ and the restriction $f|_{\Omega_i}$ to each set Ω_i is Lipschitz (see the proof of Theorem A.31). It is well-known that the Area formula holds for Lipschitz mappings and hence on each Ω_i we get

$$\int_{\Omega_i} \eta(f(x)) \, |J_f(x)| \, dx = \int_{\mathbf{R}^n} \eta(y) \, N(f, \Omega_i, y) \, dy \, .$$

By summing of these equalities we obtain the left-hand side of (A.18) since $|S| = 0$. The right-hand is bigger or equal since

$$\int_{\mathbf{R}^n} \eta(y) \, N(f, \Omega, y) \, dy = \int_{\mathbf{R}^n} \eta(y) \, N(f, S, y) \, dy + \sum_{i=1}^\infty \int_{\mathbf{R}^n} \eta(y) \, N(f, \Omega_i, y) \, dy$$

and the first term is nonnegative and potentially strictly positive if $|f(S)| > 0$. Moreover, if f satisfies the Lusin (N) condition then $|f(S)| = 0$ and hence the first term vanishes and the equality holds in (A.18). $\qquad\square$

Corollary A.36. *(a) Let $f \in W^{1,1}_{\mathrm{loc}}(\Omega, \mathbf{R}^n)$ be a homeomorphism, $\tilde{\eta}$ a nonnegative Borel measurable function on $\mathbf{R}^n$ and let $A \subset \Omega$ be a Borel measurable set. Then*

$$\int_A \tilde{\eta}(f(x)) \, |J_f(x)| \, dx \le \int_{f(A)} \tilde{\eta}(y) \, dy \, . \qquad (A.19)$$

(b) *Let* $f \in W^{1,1}_{\text{loc}}(\Omega, \mathbf{R}^n)$ *be a homeomorphism and let* η *be a nonnegative Borel measurable function on* $\mathbf{R}^n$. *Then there is a set* $\Omega' \subset \Omega$ *of full measure* $|\Omega'| = |\Omega|$ *such that*

$$\int_{\Omega'} \eta(f(x)) |J_f(x)| \, dx = \int_{f(\Omega')} \eta(y) \, dy \, . \tag{A.20}$$

(c) *Let* $f \in W^{1,1}_{\text{loc}}(\Omega, \mathbf{R}^n)$ *be a homeomorphism, let* η *be a nonnegative Borel measurable function on* $\mathbf{R}^n$ *and let* A *denote the set where* f *is differentiable. Then*

$$\int_A \eta(f(x)) |J_f(x)| \, dx = \int_{f(A)} \eta(y) \, dy \, .$$

(d) *Let* $f \in W^{1,1}_{\text{loc}}(\Omega, \mathbf{R}^n)$ *be a mapping whose multiplicity is essentially bounded by* N *and let* $A \subset \Omega$ *be a measurable set. Then*

$$\int_A |J_f(x)| \, dx \leq N \int_{f(A)} dy = N |f(A)| \, .$$

Especially we get that the Jacobian of a mapping with essentially bounded multiplicity is locally integrable.

For the part (a) above, the multiplicity of a homeomorphism is bounded by one and we apply the previous theorem for $\tilde{\eta} = \chi_{f(A)}\eta$. Regarding (b), it is enough to set $\Omega' = \bigcup_{n=1}^{\infty} \Omega_i$, where the sets Ω_i are as in the proof of the previous theorem. The part (c) follows from the previous theorem and the fact that the Lusin condition (N) holds on the set A of differentiability. This can be easily shown from the definition of differentiability with the help of the Vitali covering theorem.

The Sard Theorem [90, Theorem 7.6] tells that the image of the set of critical points is of measure zero.

Theorem A.37 (Sard). *Let* $f : \mathbf{R}^n \to \mathbf{R}^n$ *be Lipschitz mapping. Then*

$$\mathscr{L}_n(\{f(x) : J_f(x) = 0\}) = 0 \, .$$

The coarea formula is very useful tool in analysis and its form for Lipschitz functions can be found in Federer [31, 3.2.12].

Theorem A.38. *Let* $\Omega \subset \mathbf{R}^n$ *be an open set and let* $f : \Omega \to \mathbf{R}^m$ *be Lipschitz. Then for every measurable set* $E \subset \Omega$ *we have*

$$\int_E |J_m f(x)| \, dx = \int_{\mathbf{R}^m} \mathscr{H}^{n-m}(E \cap f^{-1}(y)) \, dy$$

where $J_m f$ *denotes the square root of the sum of the squares of the determinants of the m-by-m minors of the differential of* f.

A.9   Estimates for q-Capacity

The following estimate on q-capacity is standard, see e.g. [58, Theorem 5.9].

Theorem A.39. *Let $n - 1 \leq q < n$ and $\varepsilon > 0$. Let $B_0 \subset \mathbf{R}^n$ be a ball of radius r, $n \geq 2$, and let $E, F \subset \frac{1}{2}B_0$. Suppose that $u \in W^{1,q}(B_0)$ is continuous and satisfies $u \leq 0$ on E and $u \geq 1$ on F. Then*

$$r^\varepsilon \int_{B_0} |Du|^q \geq C \min\{\mathcal{H}^{n-q+\varepsilon}_\infty(E), \mathcal{H}^{n-q+\varepsilon}_\infty(F)\} .$$

Moreover, for $n = 2$ and $q = 1$ this estimate is valid also for $\varepsilon = 0$.

Proof. Without loss of generality we may assume that $u_{B_0} \geq \frac{1}{2}$; otherwise we switch the role of E and F. For every $x \in E$ we set $B_i(x) = B(x, 2^{-i+1}r)$, $i \in \mathbf{N}$. As u is continuous we have $u_{B_i(x)} \to u(x)$ and hence

$$\frac{1}{2} \leq |u(x) - u_{B_0}| \leq \sum_{i=1}^\infty |u_{B_i(x)} - u_{B_{i-1}(x)}| .$$

Since $B_i \subset B_{i-1}$, we may estimate the last term similarly to (A.11) and by the Poincaré inequality, Theorem A.17, and Hölder's inequality we obtain

$$\frac{1}{2} \leq \sum_{i=0}^\infty \fint_{B_i(x)} |u(y) - u_{B_i(x)}| \, dy \leq C \sum_{i=0}^\infty 2^{-i} r \left(\fint_{B_i(x)} |Du(y)|^q \, dy \right)^{\frac{1}{q}} . \quad \text{(A.21)}$$

Let $\delta > 0$. We claim that for every $x \in E$ we can find $i_x \in \mathbf{N}$ such that

$$\delta r^{-\varepsilon} (2^{-i_x} r)^{n-q+\varepsilon} \leq \int_{B_{i_x}(x)} |Du(y)|^q \, dy , \quad \text{(A.22)}$$

provided δ is sufficiently small (in terms of q, n, ε). Otherwise (A.21) implies that

$$\frac{1}{2} \leq C \sum_{i=0}^\infty 2^{-i} r \left(\frac{1}{|B_i(x)|} \delta r^{-\varepsilon} (2^{-i} r)^{n-q+\varepsilon} \right)^{\frac{1}{q}} \leq C \delta^{\frac{1}{q}} \sum_{i=0}^\infty (2^{-i})^{\frac{\varepsilon}{q}} \leq C \delta^{\frac{1}{q}}$$

which gives us a contradiction for small enough $\delta > 0$. Let $\delta > 0$ be fixed and so small that the above holds. For each $x \in E$ we choose a ball $B_{i_x}(x)$ such that (A.22) holds. By the Vitali covering theorem, Theorem A.1, we choose a subcollection of pairwise balls B_k with radii r_k such that $E \subset \bigcup_k 5B_k$. Using (A.22) for B_k we now obtain the desired estimate

$$\mathcal{H}^{n-q+\varepsilon}_\infty(E) \leq \sum_k (5r_k)^{n-q+\varepsilon} \leq C \sum_k \frac{1}{\delta} r^\varepsilon \int_{B_k} |Du(y)|^q \, dy \leq C r^\varepsilon \int_{B_0} |Du(y)|^q \, dy .$$

$$\text{(A.23)}$$

The proof for $n = 2$ and $q = 1$ with $\varepsilon = 0$ is more demanding and we will not give it here. It follows from a stronger estimate for continuous functions in $W^{1,1}(\mathbf{R}^n)$ (see [1]):

$$\int_0^\infty \mathcal{H}^1_\infty(\{x \in \mathbf{R}^n : M|u(x)| > t\}) \, dt \leq C(n) \int_{\mathbf{R}^n} |Du(x)| \, dx .$$

To have a geometric idea, consider the simple situation

$$E = \{0\} \times [0, \operatorname{diam} E] \text{ and } F = \{t\} \times [0, \operatorname{diam} F] \text{ for some } t > 0 .$$

Then we can use the fundamental theorem of calculus for each $y \in [0, \min(\operatorname{diam} E, \operatorname{diam} F)]$, $u(0, y) \leq 0$ and $u(t, y) \geq 1$ to obtain

$$\int_0^t |Du(s, y)| \, ds \geq 1 .$$

By Fubini we get the desired estimate

$$\int_{B_0} |Df| \geq \min(\operatorname{diam} E, \operatorname{diam} F)$$

in this simple situation. $\square$

The following result is a consequence of the proof above.

Corollary A.40. *Let $n \geq 2$ and $n - 1 < q < n$. Let $\Omega \subset \mathbf{R}^n$ be an open set and let $F \subset \Omega$ be a continuum. Suppose that $u \in W^{1,q}(\Omega)$ is continuous, has compact support in Ω and satisfies $u \geq 1$ on F. Then*

$$C(q, n) \int_\Omega |Du|^q \geq \operatorname{diam}^{n-q}(F) .$$

Proof. We extend u outside Ω as zero. Without loss of generality assume that $u \leq 1$. We can clearly fix a ball B_1 such that the support of u is contained in B_1 and we can find a ball $B_0 \subset 2B_1$ such that $F \subset B_0$ and $\operatorname{diam} B_0 \leq 2 \operatorname{diam} F$.

In the case $u_{B_0} \leq \frac{1}{2}$ we proceed similarly to the previous proof. For each $x \in F$ we can find balls $B_i(x)$ such that $u_{B_i(x)} \to u(x) \geq 1$ and we have (A.21). We can choose ε, so that $n - q + \varepsilon = 1$. Since F is a continuum we obtain $\operatorname{diam} F \leq \mathcal{H}^1_\infty(F)$. As before we obtain (A.23) and hence

$$\operatorname{diam}^{n-q+\varepsilon} F \leq C \operatorname{diam}^\varepsilon B_0 \int_{B_0} |Du(y)|^q \, dy$$

which gives us the desired estimate as $\operatorname{diam} B_0 \leq 2 \operatorname{diam} F$.

It remains to consider the case $u_{B_0} > \frac{1}{2}$. Note that clearly $u_{2B_1} \leq \frac{1}{4}$ since u is supported in B_1. Now we can find balls $\tilde{B}_1, \tilde{B}_2, \ldots, \tilde{B}_k$ such that $\tilde{B}_1 = B_0$, $\tilde{B}_k = 2B_1$,

$$\tilde{B}_i \subset \tilde{B}_{i+1}, \ 2 \operatorname{diam} \tilde{B}_i \leq \operatorname{diam} \tilde{B}_{i+1}, \ \text{and} \ |\tilde{B}_{i+1}| < C|\tilde{B}_i| \ \text{for all} \ i \in \{1, \ldots, k-1\}.$$

Similarly to (A.21) we have

$$\frac{1}{4} \leq |u_{B_0} - u_{2B_1}| \leq \sum_{i=1}^{k} \fint_{\tilde{B}_i} |u(y) - u_{\tilde{B}_i}| \, dy \leq C \sum_{i=1}^{k} \operatorname{diam}(\tilde{B}_i) \left(\fint_{\tilde{B}_i} |Du(y)|^q \, dy \right)^{\frac{1}{q}}.$$

Since $2 \operatorname{diam} \tilde{B}_i \leq \operatorname{diam} \tilde{B}_{i+1}$ and $q < n$, this implies that

$$C \leq \sum_{i=1}^{k} \operatorname{diam}^{1-\frac{n}{q}}(\tilde{B}_i) \left(\int_{\tilde{B}_i} |Du(y)|^q \, dy \right)^{\frac{1}{q}} \leq C \operatorname{diam}^{1-\frac{n}{q}}(\tilde{B}_1) \left(\int_{\Omega} |Du(y)|^q \, dy \right)^{\frac{1}{q}},$$

which gives us the desired estimate as $\operatorname{diam} \tilde{B}_1 = \operatorname{diam} B_0 \geq \operatorname{diam} F$. □

The following capacitary estimate can be found in [34].

Theorem A.41. *Suppose that Ω is a bounded open set, $E \subset \Omega$ is a continuum and that a continuous function $u \in W^{1,n}(\Omega)$ satisfies $u \geq 1$ on E and has compact support in Ω. Then*

$$\int_{\Omega} |Du(x)|^n \, dx \geq \frac{\omega_{n-1}}{\left(\log \left(\frac{C(n) \operatorname{diam} \Omega}{\operatorname{diam} E} \right) \right)^{n-1}}.$$

On the other hand, for $\Omega = B(0, R)$ and $E = \overline{B(0, r)}$ with $0 < r < R$, the above estimate is sharp as can be shown by taking $u(x) = \min\{1, \log \frac{R}{|x|} / \log(\frac{R}{r})\}$.

A.10 Solvability of $\Delta u = \varphi$

Solvability of the Poisson equation follows by convolution with the Green's function, see [28, Chap. 2.2.1 (b)].

Theorem A.42. *Let $\varphi \in C_C(\mathbf{R}^n)$. Then there is $u \in C^2(\mathbf{R}^n)$ such that $\Delta u(x) = \varphi(x)$.*

References

1. Adams, D.R.: A note on Choquet integrals with respect to Hausdorff capacity. In: Function Spaces and Applications (Lund, 1986). Lecture Notes in Mathematics, vol. 1302, pp. 115–124. Springer, Berlin (1988)
2. Ambrosio, L., Fusco, N., Pallara, D.: Functions of Bounded Variation and Free Discontinuity Problems. Oxford Mathematical Monographs. The Clarendon Press, New York (2000)
3. Astala, K., Iwaniec, T., Koskela, P., Martin, G.: Mappings of BMO-bounded distortion. Math. Ann. **317**(4), 703–726 (2000)
4. Astala, K., Iwaniec, T., Martin, G.: Elliptic Partial Differential Equations and Quasiconformal Mappings in the Plane. Princeton Mathematical Series. Princeton University Press, Princeton (2009)
5. Astala, K., Iwaniec, T., Martin, G.: Deformations of annuli with smallest mean distortion. Arch. Ration. Mech. Anal. **195**(3), 899–921 (2010)
6. Astala, K., Iwaniec, T., Martin, G., Onninen, J.: Extremal mappings of finite distortion. Proc. Lond. Math. Soc. (3) **91**(3), 655–702 (2005)
7. Astala, K., Gill, J.T., Rohde, S., Saksman, E.: Optimal regularity for planar mappings of finite distortion. Ann. Inst. H. Poincaré Anal. Non Linéaire **27**(1), 1–19 (2010)
8. Ball, J.M.: Convexity conditions and existence theorems in nonlinear elasticity. Arch. Ration. Mech. Anal. **63**, 337–403 (1978)
9. Ball, J.M.: Global invertibility of Sobolev functions and the interpenetration of matter. Proc. R. Soc. Edinb. Sect. A **88**, 315–328 (1981)
10. Ball, J.M.: Progress and puzzles in nonlinear elasticity. In: Proceedings of Course on Poly-, Quasi- and Rank-One Convexity in Applied Mechanics, CISM, Udine, to appear
11. Ball, J.M.: Singularities and computation of minimizers for variational problems. In: DeVore, R., Iserles, A., Suli, E. (eds.) Foundations of Computational Mathematics. London Mathematical Society Lecture Note Series, vol. 284, pp. 1–20. Cambridge University Press, Cambridge (2001)
12. Björn, J.: Mappings with dilatation in Orlicz spaces. Collect. Math. **53**(3), 303–311 (2002)
13. Bruckner, A.M., Bruckner, J.B., Thompson, B.S.: Real Analysis. Prentice Hall, Englewood Cliffs (2008)
14. Campbell, D., Hencl, S.: A note on mappings of finite distortion: examples for the sharp modulus of continuity. Ann. Acad. Sci. Fenn. Math. **36**, 531–536 (2011)
15. Černý, J.: Homeomorphism with zero Jacobian: sharp integrability of the derivative. J. Math. Anal. Appl. **373**(3), 161–174 (2011)
16. Cesari, L.: Sulle transformazioni continue. Ann. Math. Pura Appl. **21**, 157–188 (1942)
17. Chernavskii, A.V.: Finite to one open mappings of manifolds. Mat. Sb. **65**, 357–369 (1964)

18. Chernavskii, A.V.: Remarks on the paper "Finite to one open mappings of manifolds". Mat. Sb. **66**, 471–472 (1965)

19. Clop, A., Herron, D.A.: Mappings with finite distortion in L_{loc}^p: modulus of continuity and compression of Hausdorff measure. Isr. J. Math. (to appear)

20. Csörnyei, M., Hencl, S., Malý, J.: Homeomorphisms in the Sobolev space $W^{1,n-1}$. J. Reine Angew. Math. **644**, 221–235 (2010)

21. Daneri, S., Pratelli, A.: Smooth approximation of bi-Lipschitz orientation-preserving homeomorphisms. Ann. Inst. H. Poincaré Anal. Non Linéaire (to appear)

22. Daneri, S., Pratelli, A.: A planar bi-Lipschitz extension theorem. Adv. Calc. Var. (to appear)

23. D'onofrio, L., Hencl, S., Schiattarella, R.: Bi-Sobolev homeomorphism with zero Jacobian almost everywhere. Calc. Var. Partial Differ. Equ. doi:10.1007/s00526-013-0669-6 (to appear)

24. D'onofrio, L., Schiattarella, R.: On the total variation for the inverse of a BV-homeomorphism. Adv. Calc. Var. **6**(3), 321–338 (2013)

25. di Gironimo, P., D'onofrio, L., Sbordone, C., Schiattarella, R.: Anisotropic Sobolev homeomorphisms. Ann. Acad. Sci. Fenn. Math. **36**, 593–602 (2011)

26. Drasin, D., Pankka, P.: Sharpness of Rickman's Picard theorem in all dimensions. Preprint (2012)

27. Eremenko, A.: Bloch radius, normal families and quasiregular mappings. Proc. Am. Math. Soc. **128**, 557–560 (2000)

28. Evans, L.C.: Partial Differential Equations. Graduate Studies in Mathematics. American Mathematical Society, Providence (1998)

29. Evans, L.C., Gariepy, R.F.: Measure Theory and Fine Properties of Functions. Studies in Advanced Mathematics. CRC Press, Boca Raton (1992)

30. Faraco, D., Koskela, P., Zhong, X.: Mappings of finite distortion: the degree of regularity. Adv. Math. **90**, 300–318 (2005)

31. Federer, H.: Geometric measure theory. Die Grundlehren der mathematischen Wissenschaften, 2nd edn., Band 153. Springer, New York (1996)

32. Franchi, B., Hajlasz, P., Koskela, P.: Definitions of Sobolev classes on metric spaces. Ann. Inst. Fourier **49**(6), 1903–1924 (1999)

33. Fusco, N., Moscariello, G., Sbordone, C.: The limit of $W^{1,1}$ homeomorphisms with finite distortion. Calc. Var. Partial Differ. Equ. **33**, 377–390 (2008)

34. Gehring, F.W.: Symmetrization of rings in space. Trans. Am. Math. Soc. **101**, 499–519 (1961)

35. Gehring, F.W., Hag, K.: The Ubiquitous Quasidisk. Mathematical Surveys and Monographs, vol. 184. American Mathematical Society, Providence (2012)

36. Giannetti, F., Iwaniec, T., Onninen, J., Verde, A.: Estimates of Jacobians by subdeterminants. J. Geom. Anal. **12**(2), 223–254 (2002)

37. Giannetti, F., Pasarelli di Napoli, A.: Bisobolev mappings with differential in Orlicz Zygmund classes. J. Math. Anal. Appl. **369**(1), 346–356 (2010)

38. Gill, J.T.: Integrability of derivatives of inverses of maps of exponentially integrable distortion in the plane. J. Math. Anal. Appl. **352**(2), 762–766 (2009)

39. Gol'dstein, V., Vodop'yanov, S.: Quasiconformal mappings and spaces of functions with generalized first derivatives. Sibirsk. Mat. Z. **17**, 515–531 (1976)

40. Greco, L.: A remark on the equality $\det Df = \operatorname{Det} Df$. Diff. Integral Equ. **6**(5), 1089–1100 (1993)

41. Guo, C.Y., Koskela, P., Takkinen, J.: Generalized quasidisks and conformality. Publ. Math. (to appear)

42. Guo, C.Y.: Generalized quasidisks and conformality II. Preprint (2013)

43. Guo, C.Y., Koskela, P.: Generalized John disks. Cent. Eur. J. Math. (to appear)

44. Gutlyanski, V., Ryazanov, V., Srebro, U., Yakubov, E.: The Beltrami Equation. A Geometric Approach. Developments in Mathematics, vol. 26. Springer, Berlin (2012)

45. Heinonen, J., Kilpeläinen, T.: BLD-mappings in $W^{2,2}$ are locally invertible. Math. Ann. **318**(2), 391–396 (2000)

46. Hencl, S.: Sharpness of the assumptions for the regularity of a homeomorphism. Mich. Math. J. **59**(3), 667–678 (2010)

47. Hencl, S.: Sobolev homeomorphism with zero Jacobian almost everywhere. J. Math. Pures Appl. **95**, 444–458 (2011)
48. Hencl, S., Koskela, P.: Mappings of finite distortion: discreteness and openness for quasi-light mappings. Ann. Inst. H. Poincaré Anal. Non Linéaire **22**, 331–342 (2005)
49. Hencl, S., Koskela, P.: Regularity of the inverse of a planar Sobolev homeomorphism. Arch. Ration. Mech. Anal **180**, 75–95 (2006)
50. Hencl, S., Koskela, P., Malý, J.: Regularity of the inverse of a Sobolev homeomorphism in space. Proc. R. Soc. Edinb. Sect. A **136**(6), 1267–1285 (2006)
51. Hencl, S., Koskela, P., Onninen, J.: Homeomorphisms of bounded variation. Arch. Ration. Mech. Anal **186**, 351–360 (2007)
52. Hencl, S., Koskela, P., Onninen, J.: A note on extremal mappings of finite distortion. Math. Res. Lett. **12**(2), 231–237 (2005)
53. Hencl, S., Malý, J.: Jacobians of Sobolev homeomorphisms. Calc. Var. Partial Differ. Equ. **38**, 233–242 (2010)
54. Hencl, S., Malý, J.: Mappings of finite distortion: Hausdorff measure of zero sets. Math. Ann. **324**, 451–464 (2002)
55. Hencl, S., Moscariello, G., Passarelli di Napoli, A., Sbordone, C.: Bi-Sobolev mappings and elliptic equations in the plane. J. Math. Anal. Appl. **355**, 22–32 (2009)
56. Hencl, S., Pratelli, A.: Diffeomorphic approximation of $W^{1,1}$ planar Sobolev homeomorphisms (in preparation)
57. Hencl, S., Rajala, K.: Optimal assumptions for discreteness. Arch. Ration. Mech. Anal **207**(3), 775–783 (2013)
58. Heinonen, J., Koskela, P.: Quasiconformal maps in metric spaces with controlled geometry. Acta Math. **181**, 1–61 (1998)
59. Herron, D.A., Koskela, P.: Mappings of finite distortion: gauge dimension of generalized quasicircles. Ill. J. Math. **47**(4), 1243–1259 (2003)
60. Iwaniec, T., Koskela, P., Martin, G., Sbordone, C.: Mappings of finite distortion: $L^n \log^{alpha} L$-integrability. J. Lond. Math. Soc. **67**, 123–136 (2003)
61. Iwaniec, T., Koskela, P., Martin, G.: Mappings of BMO-distortion and Beltrami-type operators. J. Anal. Math. **88**, 337–381 (2002)
62. Iwaniec, T., Koskela, P., Onninen, J.: Mappings of finite distortion: monotonicity and continuity. Invent. Math. **144**, 507–531 (2001)
63. Iwaniec, T., Koskela, P., Onninen, J.: Mappings of finite distortion: compactness. Ann. Acad. Sci. Fenn. Math. **27**, 391–417 (2002)
64. Iwaniec, T., Kovalev, L.V., Koh, N.-T., Onninen, J.: Existence of energy-minimal diffeomorphisms between doubly connected domains. Invent. Math. **186**(3), 667–707 (2011)
65. Iwaniec, T., Kovalev, L.V., Onninen, J.: Diffeomorphic approximation of Sobolev homeomorphisms. Arch. Ration. Mech. Anal. **201**(3), 1047–1067 (2011)
66. Iwaniec, T., Kovalev, L.V., Onninen, J.: Hopf differentials and smoothing Sobolev homeomorphisms. Int. Math. Res. Not. IMRN **214**, 3256–3277 (2012)
67. Iwaniec, T., Martin, G.: Geometric Function Theory and Nonlinear Analysis. Oxford Mathematical Monographs. Clarendon Press, Oxford (2001)
68. Iwaniec, T., Onninen, J.: n-Harmonic mappings between annuli. Mem. Am. Math. Soc. **218**, 105 pp. (2012)
69. Iwaniec, T., Sbordone, C.: On the integrability of the Jacobian under minimal hypotheses. Arch. Ration. Mech. Anal. **119**, 129–143 (1992)
70. Iwaniec, T., Šverák, V.: On mappings with integrable dilatation. Proc. Am. Math. Soc. **118**, 181–188 (1993)
71. Kauhanen, J., Koskela, P., Malý, J.: Mappings of finite distortion: condition N. Mich. Math. J. **49**, 169–181 (2001)
72. Kauhanen, J., Koskela, P., Malý, J.: Mappings of finite distortion: discreteness and openness. Arch. Ration. Mech. Anal. **160**, 135–151 (2001)
73. Kauhanen, J., Koskela, P., Malý, J., Onninen, J., Zhong, X.: Mappings of finite distortion: sharp Orlicz-conditions. Rev. Mat. Iberoamericana **19**, 857–872 (2003)

74. Kleprlík, L.: Mappings of finite signed distortion: Sobolev spaces and composition of mappings. J. Math. Anal. Appl. **386**, 870–881 (2012)
75. Koskela, P.: Lectures on Quasiconformal and Quasisymmetric Mappings. University of Jyväskylä (to appear)
76. Koskela, P., Malý, J.: Mappings of finite distortion: the zero set of the Jacobian. J. Eur. Math. Soc. **5**, 95–105 (2003)
77. Koskela, P., Malý, J., Zürcher, T.: Luzin's Condition (N) and Modulus of Continuity. (2013) arXiv:1309.3094
78. Koskela, P., Onninen, J.: Mappings of finite distortion: capacity and modulus inequalities. J. Reine Angew. Math. **599**, 1–26 (2006)
79. Koskela, P., Onninen, J., Rajala, K.: Mappings of finite distortion: decay of the Jacobian. J. Geom. Anal. **22**(4), 964–976 (2012)
80. Koskela, P., Takkinen, J.: Mappings of finite distortion: formation of cusps. Publ. Mat. **51**(1), 223–242 (2007)
81. Koskela, P., Takkinen, J.: Mappings of finite distortion: formation of cusps. III. Acta Math. Sin. (Engl. Ser.) **26**(5), 817–824 (2010)
82. Koskela, P., Takkinen, J.: A note to "Mappings of finite distortion: formation of cusps II". Conform. Geom. Dyn. **14**, 184–189 (2010)
83. Koskela, P., Zapadinskaya, A., Zürcher, T.: Mappings of finite distortion: generalized Hausdorff dimension distortion. J. Geom. Anal. **20**(3), 690–704 (2010)
84. Koskela, P., Zapadinskaya, A.: Generalized dimension estimates for images of porous sets under monotone Sobolev mappings. Proc. Am. Math. Soc. (to appear)
85. Kovalev, L.V., Onninen, J.: On invertibility of Sobolev mappings. J. Reine Angew. Math. **656**, 1–16 (2011)
86. Kovalev, L.V., Onninen, J., Rajala, K.: Invertibility of Sobolev mappings under minimal hypotheses. Ann. Inst. H. Poincaré Anal. Non Linéaire **27**(2), 517–528 (2010)
87. Lukeš, J., Malý, J.: Measure and Integral. Matfyzpress, Prague (2005)
88. Malý, J.: A simple proof of the Stepanov theorem on differentiability almost everywhere. Expo. Math. **17**, 59–61 (1999)
89. Malý, J., Martio, O.: Lusin's condition (N) and mappings of the class $W^{1,n}$. J. Reine Angew. Math. **458**, 19–36 (1995)
90. Manfredi, J.: Weakly monotone functions. J. Geom. Anal. **4**, 393–402 (1994)
91. Manfredi, J., Villamor, E.: An extension of Reshetnyak's theorem. Indiana Univ. Math. J. **47**, 1131–1145 (1998)
92. Marcus, M., Mizel, V.J.: Transformations by functions in Sobolev spaces and lower semicontinuity for parametric variational problems. Bull. Am. Math. Soc. **79**, 790–795 (1973)
93. McKubre-Jordens, M., Martin, G.: Deformations with smallest weighted L^p average distortion and Nitsche type phenomena. J. Lond. Math. Soc. **85**(2), 282–300 (2012)
94. Martio, O., Ryazanov, V., Srebro, U., Yakubov, E.: Moduli in Modern Mapping Theory. Springer, Berlin (2009)
95. Mattila, P.: Geometry of Sets and Measures in Euclidean Spaces. Fractals and Rectifiability. Cambridge Studies in Advanced Mathematics, vol. 44. Cambridge University Press, Cambridge (1995)
96. Mora-Corral, C.: Approximation by piecewise affine homeomorphisms of Sobolev homeomorphisms that are smooth outside a point. Houst. J. Math. **35**, 515–539 (2009)
97. Mora-Corral, C., Pratelli, A.: Approximation of piecewise affine homeomorphisms by diffeomorphisms. J. Geom. Anal. doi:10.1007/s12220-012-9378-1 (to appear)
98. Moscariello, G., Pasarelli di Napoli, A.: The regularity of the inverses of Sobolev homeomorphisms with finite distortion. J. Geom. Anal. doi:10.1007/s12220-012-9345-x (to appear)
99. Müller, S.: Higher integrability of determinants and weak convergence in L^1. J. Reine Angew. Math. **412**, 20–34 (1990)
100. Onninen, J.: Mappings of finite distortion: minors of the differential matrix. Calc. Var. Partial Differ. Equ. **21**, 335–348 (2004)

101. Onninen, J., Zhong, X.: A note on mappings of finite distortion: the sharp modulus of continuity. Mich. Math. J. **53**(2), 329–335 (2005)
102. Onninen, J., Zhong, X.: Mappings of finite distortion: a new proof for discreteness and openness. Proc. R. Soc. Edinb. Sect. A **138**, 1097–1102 (2008)
103. Palka, B.P.: An Introduction to Complex Function Theory. Undergraduate Texts in Mathematics. Springer, Berlin (1995)
104. Pasarelli di Napoli, A.: Bisobolev mappings and homeomorphisms with finite distortion. Rend. Lincei Mat. Appl., **23**(4), 1–18 (2012)
105. Ponomarev, S.: Examples of homeomorphisms in the class $ACTL^p$ which do not satisfy the absolute continuity condition of Banach (Russian). Dokl. Akad. Nauk USSR **201**, 1053–1054 (1971)
106. Pratelli, A., Puglisi, S.: Elastic deformations on the plane and approximations. Preprint (2013)
107. Rado, T., Reichelderfer, P.V.: Continuous Transformations in Analysis. Springer, Berlin (1955)
108. Rajala, K.: Mappings of finite distortion: the Rickman-Picard theorem for mappings of finite lower order. J. Anal. Math. **94**, 235–248 (2004)
109. Rajala, K.: A lower bound for the Bloch radius of K-quasiregular mappings. Proc. Am. Math. Soc. **132**, 2593–2601 (2004)
110. Rajala, K.: Bloch's theorem for mappings of bounded and finite distortion. Math. Ann. **339**, 445–460 (2007)
111. Rajala, K.: Remarks on the Iwaniec-Šverák conjecture. Indiana Univ. Math. J. **59**, 2027–2040 (2010)
112. Rajala, K.: Reshetnyak's theorem and the inner distortion. Pure Appl. Math. Q. **7**, 411–424 (2011)
113. Rajala, T.: Planar Sobolev homeomorphisms and Hausdorff dimension distortion. Proc. Am. Math. Soc. **139**(5), 1825–1829 (2011)
114. Reshetnyak, Yu.G.: Certain geometric properties of functions and mappings with generalized derivatives. Sibirsk. Mat. Z. **7**, 886–919 (1966)
115. Reshetnyak, Yu.G.: Space mappings with bounded distortion. Sibirsk. Mat. Z. **8**, 629–658 (1967)
116. Reshetnyak, Yu.G.: Space Mappings with Bounded Distortion. Translations of Mathematical Monographs, vol. 73. American Mathematical Society, Providence (1989)
117. Rickman, S.: Quasiregular Mappings. Ergebnisse der Mathematik und ihrer Grenzgebiete (3) (Results in Mathematics and Related Areas (3)), vol. 26. Springer, Berlin (1993)
118. Šverák, V.: Regularity properties of deformations with finite energy. Arch. Rational Mech. Anal. **100**, 105–127 (1988)
119. Tengvall, V.: Differentiability in the Sobolev space $W^{1,n-1}$. Calc. Var. Partial Differ. Equ. (to appear)
120. Ukhlov, A.D.: On mappings generating the embeddings of Sobolev spaces. Siberian Math. J. **34**, 165–171 (1993)
121. Zapadinskaya, A.: Generalized dimension compression under mappings of exponentially integrable distortion. Cent. Eur. J. Math. **9**(2), 356–363 (2011)
122. Ziemer, W.P.: Weakly Differentiable Function: Sobolev Spaces and Functions of Bounded Variation. Graduate Text in Mathematics, vol. 120. Springer, New York (1989)

Index

S. Hencl and P. Koskela, *Lectures on Mappings of Finite Distortion*, Lecture Notes
in Mathematics 2096, DOI 10.1007/978-3-319-03173-6,
© Springer International Publishing Switzerland 2014

LECTURE NOTES IN MATHEMATICS  Springer

Edited by J.-M. Morel, B. Teissier; P.K. Maini

Editorial Policy (for the publication of monographs)

1. Lecture Notes aim to report new developments in all areas of mathematics and their applications - quickly, informally and at a high level. Mathematical texts analysing new developments in modelling and numerical simulation are welcome.

   Monograph manuscripts should be reasonably self-contained and rounded off. Thus they may, and often will, present not only results of the author but also related work by other people. They may be based on specialised lecture courses. Furthermore, the manuscripts should provide sufficient motivation, examples and applications. This clearly distinguishes Lecture Notes from journal articles or technical reports which normally are very concise. Articles intended for a journal but too long to be accepted by most journals, usually do not have this "lecture notes" character. For similar reasons it is unusual for doctoral theses to be accepted for the Lecture Notes series, though habilitation theses may be appropriate.

2. Manuscripts should be submitted either online at www.editorialmanager.com/lnm to Springer's mathematics editorial in Heidelberg, or to one of the series editors. In general, manuscripts will be sent out to 2 external referees for evaluation. If a decision cannot yet be reached on the basis of the first 2 reports, further referees may be contacted: The author will be informed of this. A final decision to publish can be made only on the basis of the complete manuscript, however a refereeing process leading to a preliminary decision can be based on a pre-final or incomplete manuscript. The strict minimum amount of material that will be considered should include a detailed outline describing the planned contents of each chapter, a bibliography and several sample chapters.

   Authors should be aware that incomplete or insufficiently close to final manuscripts almost always result in longer refereeing times and nevertheless unclear referees' recommendations, making further refereeing of a final draft necessary.

   Authors should also be aware that parallel submission of their manuscript to another publisher while under consideration for LNM will in general lead to immediate rejection.

3. Manuscripts should in general be submitted in English. Final manuscripts should contain at least 100 pages of mathematical text and should always include

   – a table of contents;
   – an informative introduction, with adequate motivation and perhaps some historical remarks: it should be accessible to a reader not intimately familiar with the topic treated;
   – a subject index: as a rule this is genuinely helpful for the reader.

   For evaluation purposes, manuscripts may be submitted in print or electronic form (print form is still preferred by most referees), in the latter case preferably as pdf- or zipped ps-files. Lecture Notes volumes are, as a rule, printed digitally from the authors' files. To ensure best results, authors are asked to use the LaTeX2e style files available from Springer's web-server at:

   ftp://ftp.springer.de/pub/tex/latex/svmonot1/ (for monographs) and
   ftp://ftp.springer.de/pub/tex/latex/svmultt1/ (for summer schools/tutorials).

Additional technical instructions, if necessary, are available on request from lnm@springer.com.

4. Careful preparation of the manuscripts will help keep production time short besides ensuring satisfactory appearance of the finished book in print and online. After acceptance of the manuscript authors will be asked to prepare the final LaTeX source files and also the corresponding dvi-, pdf- or zipped ps-file. The LaTeX source files are essential for producing the full-text online version of the book (see http://www.springerlink.com/openurl.asp?genre=journal&issn=0075-8434 for the existing online volumes of LNM). The actual production of a Lecture Notes volume takes approximately 12 weeks.

5. Authors receive a total of 50 free copies of their volume, but no royalties. They are entitled to a discount of 33.3 % on the price of Springer books purchased for their personal use, if ordering directly from Springer.

6. Commitment to publish is made by letter of intent rather than by signing a formal contract. Springer-Verlag secures the copyright for each volume. Authors are free to reuse material contained in their LNM volumes in later publications: a brief written (or e-mail) request for formal permission is sufficient.

Addresses:
Professor J.-M. Morel, CMLA,
École Normale Supérieure de Cachan,
61 Avenue du Président Wilson, 94235 Cachan Cedex, France
E-mail: morel@cmla.ens-cachan.fr

Professor B. Teissier, Institut Mathématique de Jussieu,
UMR 7586 du CNRS, Équipe "Géométrie et Dynamique",
175 rue du Chevaleret
75013 Paris, France
E-mail: teissier@math.jussieu.fr

For the "Mathematical Biosciences Subseries" of LNM:

Professor P. K. Maini, Center for Mathematical Biology,
Mathematical Institute, 24-29 St Giles,
Oxford OX1 3LP, UK
E-mail: maini@maths.ox.ac.uk

Springer, Mathematics Editorial, Tiergartenstr. 17,
69121 Heidelberg, Germany,
Tel.: +49 (6221) 4876-8259

Fax: +49 (6221) 4876-8259
E-mail: lnm@springer.com